THE ORIGINS OF HUMAN POTENTIAL

Few scientific debates have been more intense and protracted than the 'nature or nurture' debate about the origins of human potential. *The Origins of Human Potential* revisits the 'nature–nurture' of cognitive ability debate in the light of new evidence from evolutionary biology. In the last few years new claims have been made about the biological structure of our knowledge, of ways in which we think, and of individual differences in them. When such debates persist around scientific topics of serious social concern it is often the case that they are based, not on 'pure' facts, but on deep-seated assumptions that may be false. This book identifies and examines the host of assumptions that enter into nature–nurture arguments in psychology, and shows how they have retarded understanding and perpetuated sterility of argument.

Among the key assumptions examined are those about the nature of genes and their involvement in development; the nature of evolution as it has impinged on cognitive ability; and the nature of cognitive ability itself, especially its reduction to a simple 'quantitative' characteristic like height or weight alleged to be measured by IQ tests. Assumptions underlying IQ testing are examined, as are those underlying the famous twin and adoption studies said to show that the ladder of cognitive ability 'revealed' by IQ tests is caused by genetic differences. What emerges is the development of a child's *socio*-cognitive abilities, with a much greater role for self-organised, creative individuals than hitherto recognised.

Raising complex issues in a clear way, *The Origins of Human Potential* also provides new insights into the contentious area of educational policy. It will be of great relevance to students of psychology and education, and should be read by all those with an interest in our children's future.

Ken Richardson is Senior Lecturer in the Centre for Human Development and Learning at the Open University. His previous publications include *Understanding Psychology* (1989) and *Understanding Intelligence* (1991).

THE ORIGINS OF HUMAN POTENTIAL

Evolution, development and psychology

Ken Richardson

London and New York

First published 1998
by Routledge
11 New Fetter Lane, London EC4P 4EE

Simultaneously published in the USA and Canada
by Routledge
29 West 35th Street, New York, NY 10001

Typeset in Times by J&L Composition Ltd, Filey, North Yorkshire
Printed and bound in Great Britain by TJ International Ltd, Padstow, Cornwall

British Library Cataloguing in Publication Data
A catalogue record for this book is available from the British Library

Library of Congress Cataloging-in-Publication Data
Richardson, Ken.
The origins of human potential: evolution, development and psychology/Ken Richardson.
p. cm.
Includes bibliographical references and index.
1. Intellect. 2. Nature and nurture. I. Title.
BF431.R414 1998
97–14925
CIP
153.9–dc21

ISBN 0–415–17369–8 (hbk)
ISBN 0–415–17370–1 (pbk)

CONTENTS

FIGURES

FIGURES

PREFACE

I hope that only a small amount of thought on the part of the reader will dispel any ambiguity in the title of this book. There seems little doubt that when we consider specifically human potential we are most likely to think, first and foremost (though certainly not exclusively) of human cognitive ability – and that, indeed, is what this book is about. There is little mystery about why human *cognitive* ability has always been at the top of the agenda in psychology. Knowing and reasoning, and the abilities they furnish, almost define the human species. Remarkably, though, psychological theory about what cognitive ability is, and where it comes from, remains, after two thousand years of scholarship, in a very backward state. The origins of human potential in this sphere thus remain the grounds around which many heated debates continue, as every psychologist knows.

There are probably many reasons for this theoretical backwardness. Prominent among them must be the kinds of assumptions which psychologists adopt as a basis for their research and theorising. When endless debates persist around important questions it is usually the case that there is something wrong with the assumptions around which the questions arise, and thus the evidential material which is used to address them. As the philosopher R.G. Collingwood said many years ago, every scientific statement is the answer to a question, but every question is based on a presupposition.

Even a cursory glance at theories and statements in the area in question will demonstrate the surfeit of assumptions – about the nature of potential, its origins, what it is potential *in* or *for*, and so on. This book is about such assumptions. The evaluation of assumptions is itself a worthwhile task because it helps put students, researchers, and others, in a critical position with respect to the ideas they deal with, instead of blindly accepting 'facts'. In turn, such criticism helps provide the basis for new knowledge. This work is not about *all* such assumptions: rather, I confine myself to those which seem to predominate in current thinking. As the Russian psychologist L.S. Vygotsky noted long ago (see Van der Veer

1996), because of the lack of any general theoretical system in psychology, some assumptions can become disproportionately potent and grow into 'general philosophical worldviews' that dominate the views of scientists, politicians, and the general public. One example earlier this century was the concept of the unconscious. Although originally constructed to explain the nature of neuroses, it grew into a general theory of personality, before finally being used to explain all human creativity, human strife and even warfare (Van der Veer 1996).

A similar process seems to have happened in psychology with the notion of the gene. Although notions of innate potential in cognitive ability (and of innate individual differences in it) pre-date that of the gene by over two thousand years, the appeal of the gene as the agent of such potential has grown (with some ups and downs) throughout this century until it has come to dominate thinking in recent times. Everyone knows that there are actually two 'gene stories' involved in this dominance. The first is that the form or structure of cognitive ability has been fashioned by natural selection and is now encoded in our genes (just like, so the story goes, our more visible physical characters such as the hand and eye). One of the most alarming consequences of this story is that psychology students and theorists are now everywhere being urged to become amateur Darwinists. 'Psycho-Darwinism', as I've heard it called, has become the latest 'explain all' idea.

A related argument – and this is the second story – is that Darwinian principles suggest significant genetic variability underlying, and determining individual differences in, cognitive potential in human populations. This story, too, has created many problems, and just as many debates. Based on rather simple 'measures' of potential, it is a story which has powerfully captured the minds of psychologists and the general public alike, far beyond any strictly scientific purpose. Thus, we all know how it has repeatedly been taken to explain a rank ordering of society, the relative abilities of different social classes and sexes, as well as periodically erupting in the spectacle of scientific racism.

The main problem with both stories, I shall argue, is that they have involved the erection of a very special, idealistic 'gene', rather different from the gene we read about in the developmental and molecular genetics literature. This idealistic 'gene' is then used to explain so much about cognition and its diversity, until it has become, once again, a 'general philosophical worldview'. Both stories, of course, also involve rather vague and idealistic notions of 'environments', of 'interactions', and cognitive ability itself (as in the pure abstraction known as IQ), which I also come to criticise. In addition, though, I want to show how such idealism draws self-contradictions from its authors: in my view, these together do much to explain the continuing backwardness and fragmentation in the discipline generally.

As an antidote, I resort to a 'systems' view of development, showing how human cognitive (or, more accurately, *socio*-cognitive) abilities are the evolutionary tip of new regulatory mechanisms, involving developmental processes and hierarchical interrelations using genetic materials as developmental resources. This will entail some novel psychology; but it also involves a glimpse at the revolutionary advances in evolutionary biology and developmental genetics in recent years. These advances are amply justifying Darwin's own circumspection, as expressed with some exasperation in his Preface to the last edition of *Origin of Species* (1872):

> But, as my conclusions have lately been much misrepresented . . . I may be permitted to remark that in the first edition of this work, and subsequently, I placed in a most conspicuous position – namely at the close of the Introduction – the following words: 'I am convinced that natural selection has been the main, but not the exclusive means of modification'. This has been of no avail. Great is the power of steady misrepresentation.
>
> (Quoted by Bocchi 1991: 42-43)

It is those other 'means of modification' that cognitive ability (or, more accurately, *socio*-cognitive ability) seems most spectacularly to exemplify. I hope to show how such a view is not only a more realistic explanation of adaptability in a dynamically changing world, especially a human social world, but also one that requires us to dump a great deal of archaic conceptual furniture in psychology. Evolutionary and developmental biologists have done much in recent years to reveal the more general importance of those other means. The current irony is that, while psychologists are increasingly seeking to anchor their unstable ideas in Darwinian biology, evolutionary biologists are moving into a 'post-Darwinian' era.

Of course, one of the things I am trying to do in this work is to change the basic terms of nature–nurture debates in psychology – but not by the usual obfuscations of talking about 'interactions', as if consisting of a vague admixture of genes and environments. I do not accept the ancient assumptions on which such accounts are based, either as they refer to the nature of cognitive characters, to genes or to environments, so I do not entertain the classic questions which arise from them. The reader who is seeking answers to questions like 'Which are most important, genes or environments?', or 'To what extent are cognitive abilities determined by genes?', will be disappointed, therefore, because I do not accept the presuppositions on which such questions are based. Psychologists who want to see a role for genes, *will* see one here – though it may not be the one popular in psychology since the turn of the century.

Similarly for those who want to stress the role of the 'environment'. What all will see is a more strongly asserted role for development itself, and for self-organising, creative beings at the centre of things. Because I do not seek a traditionally 'safe' or 'balanced' middle ground on the subject, I have little doubt that I will be accused of adopting an extreme position by the superficial reader looking for just that. I can only suggest to such readers that they read the book again!

The current importance of the subject of cognitive ability probably goes without saying. All over the Western world the cognitive ability of individuals has become increasingly valued for its supposed pay-off in 'the current economic climate', and the pressures in education systems to find it have increased accordingly. I hope to show how such a mechanistic view of cognitive ability, resting on some 'innate potential' of individuals, is self-defeating, and accounts in no small way for the constant problems in our education systems. But there is now another source of urgency of the sort of critique offered here. As Michael Cole (1990: 107) notes, the same imperatives about cognitive ability are being imposed on less-developed countries worldwide, 'inserted into the social process by powerful outsiders', as a 'passkey' to Western economics, but also as 'a source of social disruption and human misery'. The danger is, then, that the same explanatory framework, now prevalent in the West, will come to encompass all the world's children. Only a radically different view of the nature of cognitive ability can help prevent this, and this is what I have tried to construct.

One of the most exciting aspects of the world of ideas (and one which itself, in my view, belies the biological determinism now prevalent) is that of working creatively with the ideas of numerous other people. My attempt to reassert human cognitive ability as a distinct level of process, and thus of scientific explanation, builds upon a number of 'hierarchical' accounts in recent years, both in evolutionary biology (see e.g. Depew and Weber 1995; Plotkin 1994; Nederbragt 1997) and in psychology and the brain sciences (Rose 1981; Rose, Kamin, and Lewontin 1984; Oyama 1985). I can only hope that I have added something to the work of those conceptual pathbreakers that will be of interest and use to psychologists, and indeed all of those who are interested in the future of the world's children.

I want to thank all friends and colleagues who read, and commented on, an earlier draft of this work. I would particularly like to thank Steven Rose for specific helpful suggestions and for the germs, at least, of many of the ideas presented in this book, sown in the distant past. David Webster also set me straight on a few important points which I am sure has been beneficial. For their sake, I have to admit, as usual, that the remaining defects are entirely my own. I want to thank Margaret van Burken for doing so much to maintain order in my Departmental business in spite of frequent distractions. Finally, I would like to thank my partner Susan (the best 'crap-detector' I have ever known) for her unswerving encouragement in having me tackle what she has always known to be seriously muddled – and seriously consequential – issues.

References

Bocchi, G. (1991). Biological evolution: the changing image. In E. Laszlo (ed.) *The New Evolutionary Paradigm.* London: Gordon & Breach.

Cole, M. (1990). Cognitive development and formal schooling: the evidence from cross-cultural research. In L.C. Moll (ed.) *Vygotsky and Education: Instructional Implications and Applications of Sociohistorical Psychology.* Cambridge: Cambridge University Press.

Depew, D.J. and Weber, B.H. (1995). *Darwinism Evolving.* Cambridge, MA: MIT Press.

Nederbragt, H. (1997). Hierarchical organisation of biological systems and the structure of adaptation in evolution and tumorigenesis. *Journal of Theoretical Biology*, 184, 149–156.

Oyama, S. (1985). *The Ontogeny of Information.* Cambridge: Cambridge University Press.

Plotkin, H.C. (1994). *The Nature of Knowledge.* London: Penguin.

Rose, S. (1981). From causations to translations: what biochemistry can contribute to the study of behaviour. In P.P.G. Bateson and P.H. Klopfer (eds) *Perspectives in Ethology*, Vol. 4. New York: Plenum Press.

Rose, S., Kamin, L.J., and Lewontin, R.C. (1984). *Not in Our Genes.* Harmondsworth: Penguin.

Van der Veer, R. (1996). Structure and development. Reflections by Vygotsky. In A. Tryphon and J. Voneche (eds) *Piaget–Vygotsky.* Hove: Psychology Press.

1

GENE GODS

Introduction: nativism and mystery

There is nothing on earth even remotely like human cognitive ability. Apparent in some other animals, cognitive potential becomes obvious even in early infancy in humans, bursts into prominence in toddlers and preschoolers, becomes laboriously harnessed in schools, and then continues on a course of lifelong development, producing all the unique adaptability and creativity with which human life and human history are littered. So dominant is cognitive ability in humans, in fact, that it emphatically defines the species. It is hardly surprising that it is among the most long-standing, and simultaneously the most difficult and demanding, areas of all scientific enquiry.

Progress in *understanding* cognitive ability, however, has been neither smooth nor definite. In spite of scholarly inquiry stretching back over two thousand years, and intensive empirical and theoretical effort in modern times, its key aspects – knowing and reasoning, their origins, and diversity in them – remain obscure. This backwardness presents something of a blot on the scientific landscape. Yet so important is cognitive ability in humans, both socially and economically, that psychologists are under constant pressure to say something definite and authoritative about it – in particular how its potential may be identified, promoted, harnessed, located in particular individuals, and so on. In such circumstances, what society usually gets in reply is a mixture of empirical facts and assumptions – couched in the most persuasive terms possible!

Assumptions, of course, are used as 'straw' to substitute for empirical facts in making up a theoretical model, and they tend to come in two forms. These are the metaphysical (or just plain mystical!) and the metaphorical. The former have been recruited throughout the history of science, in all fields, sometimes without regard for empirical details at all. In his *Immortality of the Soul* (1512), Thomas More described his concept of the 'spirit of nature', a force 'immaterial and incorporeal', inscrutable in itself, yet somehow driving the processes of natural and human behaviour. As Burtt (1932: 133) pointed out, this idea was quite common in the Middle Ages, 'being appealed to frequently by mystics, theosophists and speculative natural philosophers', and stretches back to the Platonic notion of an *anima mundi*, a living principle 'which penetrates matter and whose active powers are expressed in the larger astronomical and physical phenomena of nature'. Likewise, for centuries, the Ptolemaic view of the earth-centred universe was a mix of the scholarly and the theological, disease was caused by 'bad spirits', and even a basic phenomenon like combustion was caused by mysterious 'essences' within the things that burned.

Metaphoric assumptions, on the other hand, draw explanatory power through comparison with other domains, already described and understood, or from everyday physical and social experience. Such assumptions have often been the bases of great breakthroughs by scientists; but, just as often, their downfall. So we once had theories of the universe based on the 'obvious' fact that all the heavens rotate around *us*, and of navigation based on the perception of a flat earth. Theories of cognition have traditionally been steeped in such metaphoric assumptions. So the mind has been likened to a telephone exchange, a machine, a computer, or other 'logical' operator, and falling into social class, gender, and other ranks according to the layout of people in social experience.

Indeed, the most popular notion of potential (or, more strictly, potentiality) has been largely based on metaphors of origins in, and growth or manifestation from, predetermined form. Much of this book is about such an assumption, but it has been around a long time. For example, the bursting and unfolding of buds and blossoms suggest that complex abilities like cognition might be 'inside us' all along, albeit in smaller form, and get passed on to subsequent generations in that form, and just, as it were, 'grow' in individuals. Once taken on board, such an assumption has supported major theories of origins of human potential. So the early microscopists were quick to 'see', and to show us in their drawings, a little man – a 'homunculus' – curled up in the head of a sperm. Leibniz (1646–1716) applied the same metaphor to psychology when he argued, 'the souls which will one day be human souls have been present in the semen of their ancestors down to Adam, and consequently existed since the beginning of things' (cf. Dobzhansky 1962: 25).

The developmental psychologist, Jean Piaget (e.g. 1929), has pointed out how such metaphysical and metaphorical assumptions are very common in the reasoning of young children. The preschool child habitually fails to distinguish appearances from underlying reality, and attributes causes to hidden agents as purposeful initiators of events – a naive conception of cause which Piaget called 'intentional causality'. And they tend to take appearance as literal reality, often when it involves self-contradictions. Such tendencies are only overcome, according to

Piaget, through the constructive process of cognitive development which follows in nearly all children in the course of their actions on the world. Somewhat similarly, the progress of science has been in many ways one of discrimination of underlying reality from mere appearance (Carey 1988), and of 'decentration' from mere perception as the basis of assumptions. And as Hall (1983: 4) notes, the distinctive aspect of modern science relative to antecedents is that 'it excludes spiritual agencies from its province and accepts a pure materialism'.

What I want to argue in this book is that discussion and theory surrounding the classic nature–nurture debates about cognitive ability – what it is, where it comes from, and how it is so diverse – are still full of such metaphoric assumptions, such failures to discriminate appearance from reality, and such spirit agents or mystical powers. These are manifest in any one of a number of current ideas and theories about cognition. But one idea appears to stalk the current scene above all others, and that is the nature of the gene as the roots of all things cognitive, and thus the solution to the question of origins of human potential.

Today, genes are said to determine not only bodily structures and simple reflexes, but also the fundamental form of our knowledge and cognitive processes. In addition genes are said to exist in ranks, according to their inherent powers or strengths, and thus come to determine the mental and social status of their possessors, what we can or cannot learn, and even in which school subjects children can be successes or failures. In both these ways it has been said of genes that they determine the substantive form of human social institutions, the different mental abilities of social classes, genders, and 'races', and, in a sense, the socio-economic structure of the whole human world.

All of these things have been said – and are still being said – about genes and cognitive ability. Equally remarkable and idealistic things are said about 'environmental' determinants of cognitive ability, or even about 'interactions' between genes and environments, and I shall be criticising those in various places too. But such views have been put with less clear and definite attribution, and offer weaker metaphors for the origins of potential. They have thus achieved a much lesser grasp on the minds of psychologists and the general public. 'Genetic' theories of cognitive ability, on the other hand, have presented a story of superficial clarity and simplicity, comprising powerful ideas that have flooded the century, including all our institutions, and the minds of psychologists and the general public alike. It is such ideas, I argue, that, once again, have become rather strong in recent times, yet are highly misleading and damaging, both for psychology and for people.

It is not too difficult to identify, in this story, the kinds of assumptions just mentioned. Thus, one of my aims will be to point out in this critique that the story is not precisely about the gene that is known to developmental and molecular geneticists, and whose role in the construction of characteristics I will be discussing in Chapter 2. Rather, it is an idealistic version, an image of a 'gene' powerful and monarchic, and imputed with many of the spiritual and metaphoric qualities just mentioned. As Oyama (1985) points out, the gene has itself been institutionalised as a quasi-cognitive power in psychology: an almost conscious agent, fundamentally determining both structure and diversity in cognitive ability. And this notion of the gene readily conveys the whole metaphoric package of potential, bound up in the term itself, as well as others like 'blueprints' and 'codes'.

The more notorious aspects of the career of this gene story in the twentieth century are, of course, perfectly well known. As well as promoting a 'horticultural' view of people, and a sense of immutability about social structures, the gene story became, as Rose, Kamin and Lewontin (1984: 27) note, 'the basis for the German racial and eugenic laws that began with the sterilisation of the mentally and morally undesirable and ended in Auschwitz'. But the story prevails in a myriad of seemingly more benign (but ultimately damaging) forms just as much today, among psychologists, as well as the general public. It is such prevalence in ideas of cognitive ability that this book is about.

The best indication of its influence on the minds of the public is, perhaps, the way that parents so quickly decide on the already determined abilities and personalities of their children on the basis of superficial signals. In the case of gender-related abilities, this starts from the moment of birth, when an infant's sex is deemed to be the most important thing worth knowing (Lloyd and Duveen 1991). From that moment children become attributed with a host of respective cognitive characteristics, with immediate and continual consequences for subsequent development. More generally, Newson and Newson (1977) have shown how parents feel quite sure about their judgements of their children's intelligence from an early age. Indeed, such judgements are often made in the first few days of an infant's life. Goodnow (1984) cites one study showing how this happens in the first forty-eight hours!

The best evidence of the influence of the gene story in social policy and practice has been seen in the institution of education. Indeed, for most of this century – and increasingly in recent times – the goals of the institution have been put as the cultivation of innate 'potential', and many policies, from selection at eleven years, and more recent moves to test children for their 'talents' at an early age, reflect that. Indeed, teachers' willingness to judge their pupils' innate potential has long been a basis for constructing IQ tests, as I shall describe further in Chapter 4.

From popular pictures of 'talent' and 'potential' prefigured in genes, to banks of sperm duly labelled with their donors' IQ, and the ladder view of human ranks of ability, the story permeates the public mind. The trouble is that it is just such caricatured images of the gene that have been increasingly picked up and become stuck in the minds of psychologists, in recent years, as well: indeed some psychologists have played no small role in promoting it. And it is often, if not usually, just such forms that students are expected to assume in their courses.

This first chapter is about the evidence and argument, and the assumptions underlying them, by which the *origins* and *form* of human cognitive ability have been 'found' in genes (those concerning diversity in cognitive ability being treated in Chapter 5). These assumptions include those about the nature of the genes themselves, as they are popularly conceived, and as they appear in psychology text books, i.e. as 'codes' for characters. I discuss natural selection and popular 'adaptationist' assumptions, and many other psychological outgrowths of the Darwinian framework appearing in recent years. All of these aspects I will pick up on in subsequent chapters. To start with I offer a brief review of their place in contemporary psychological theory.

Nativism in modern psychology

Although contemporary psychology is, in many ways, a backward discipline, it is worth recalling how the spiritual idealism that prevails within it, especially around the nature of genes, stems from the writings of Plato in Ancient Greece (around 340 BC). This is because much of the same reasoning about origins of cognitive abilities prevails today. Complaining that the richness, coherence, and complexity of knowledge and reasoning could not come about through accident (which is what experience or learning were deemed to be), Plato went on to 'prove' his point, in a famous Dialogue with his old friend Meno (Hamilton and Cains 1961). It is, perhaps, worth giving part of the Dialogue itself because it perfectly illustrates the evocation of spirit agents that still prevails in nature–nurture debates in psychology.

Socrates: Thus the soul which is immortal and has been born many times, and has seen all things both here and in the other world, has learned everything there is . . . so that when a man has recalled a single piece of knowledge – learned it, in ordinary language – there is no reason why he should not find out all the rest . . . for seeking and learning are in fact nothing but recollection.

Meno: I see, Socrates, but what do you mean when you say that we don't learn anything?

Socrates: It isn't an easy thing, but still I should like to do what I can since you ask me. I see you have a large number of retainers here. Call one of them, anyone you like, and I will use him to demonstrate it to you.

Meno: Certainly. [To a slave boy.] Come here.
Socrates: He is Greek and speaks our language?
Meno: Indeed yes – born and bred in the house.
Socrates: Listen carefully, then, and see whether it seems to you that he is learning from me or simply being reminded.
Meno: I will.
Socrates: Now boy, you know that a square is a figure like this.
[Socrates begins to draw a figure in the sand. He points to the square abcd]:

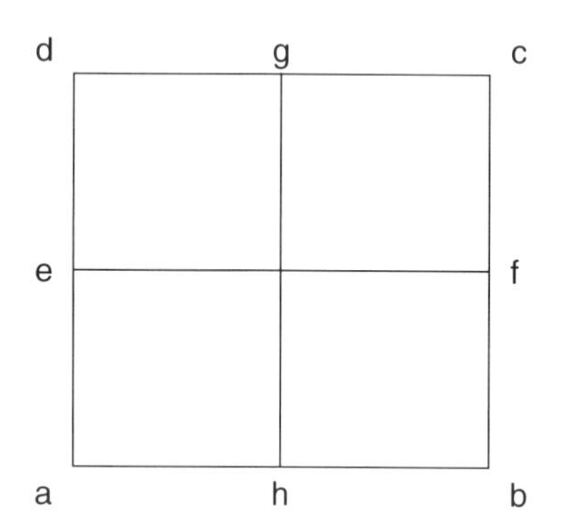

Boy: Yes.
Socrates: It has all these four sides equal?
Boy: Yes.
Socrates: And these lines which go through the middle of it are also equal?
Boy: Yes.
Socrates: Such a figure could either be larger or smaller, could it not?
Boy: Yes.
Socrates: Now . . . . If it were two feet in this direction and only one in that, must not the area be two feet taken once?
Boy: Yes.
Socrates: But since it is two feet this way also, does it not become twice two feet?
Boy: Yes.
Socrates: And how many feet is twice two? Work it out and tell me.
Boy: Four.
[And Socrates goes on, through a series of questions and drawings, ending up with the uneducated slave boy 'proving' that the square on the diagonal of a square is double the latter's area. Finally turning back to Meno]:
Socrates: What do you think Meno, has he answered with any opinions that were not his own?
Meno: No. They were all his.

Socrates: Has anyone taught him all these? You ought to know, especially as he has been brought up in your household.
Meno: I know that no one ever taught him.
Socrates: Then if he did not acquire them in this life, isn't it immediately clear that he possessed and had learned them during some other period . . . when he was not in human shape?
Meno: Yes.

(Adapted from Hamilton and Cains 1961)

Plato's theory of knowledge and reasoning was thus based on a telling question, still prominent today: how could the extraordinary richness of our knowledge and thought possibly come from personal experience, which is, he argued, bland, impoverished, fragmentary, and unreliable? Thus he suggested that all our key ideas and intellectual processes are innate, implanted within us by God at conception, and subsequently 'recalled' through processes of rational discourse (the theory became known as rationalism).

It is also well known how the Platonic mysticism was theorised in order to support another, social, purpose. If our cognitive ability is put inside us 'from the beginning' by God, the latter does not, alas, distribute it very evenly. Accordingly, Plato wrote in support of the hierarchical social-class divisions that then existed, claiming that they simply mirrored differences in innate intellectual strength among the populace (i.e. the citizens other than slaves, who were considered to be scarcely human). To improve the ability of people in the state, generally, he advocated a system of eugenic reproduction – marriage and reproduction confined to the Guardians (the Rulers). Using the inequalities of metals, from gold to iron, as an analogy, he urged a vigilance about the 'characters' of people which, in terms of superstitious fear, finds many an echo in accounts of mental ability in the twentieth century:

> Therefore the first and most important of God's commandments to the Rulers is that they must exercise their function as Guardians with particular care in watching the mixture of metals in the characters of the children. If one of their own children has bronze or iron in its make-up, they must harden their hearts, and degrade it to the ranks of the industrial and agricultural class where it properly belongs.

These conclusions, of course, suited the authoritarian regime for which Plato was writing very well (Robinson 1981). But the striking thing about the *Meno* today, apart from its cleverness, is that its basic intuitions are still widely embraced and still essentially constitute the modern gene story of cognitive ability: the idea of 'knowledge inside us', deposited by a specific formative agent 'when we are not in human shape', of already having 'learned all there is' as a result of 'being born many times', and merely 'recalled' in the current life, still form the most dominant theory of cognitive ability. Of course, in their search for legitimacy, psychologists have replaced God with the gene. But the basic causal ideas remain much the same: the action of being 'born many times' is attributed to natural selection operating across generations; knowledge is 'coded' in the genes; and 'recall' is the act of transcription from the genes into some brain and cognitive functions. And individual differences in cognitive ability are attributed to genetic differences.

According to contemporary views, then, justification for nativist assumptions about cognitive abilities is to be found in biology, and everywhere, today, psychologists are being urged to become quasi-biologists. The strength of urgency with which we are admonished to import biology into our thinking about cognitive ability is illustrated by Scarr (1993: 1334) in her Presidential Address to the American Psychological Association: 'What has been missing . . . is a full appreciation of biology, with its enormous implications for human development'. Although suggestive of an evasion of the darkness within one discipline by turning to the light of another, the assertion is also usually made with a certain edge, if not a hint of bitterness. The picture created is that of dedicated Darwinian psychologists struggling against the naive social psychologist and the politically correct: 'Behaviour geneticists . . . have fought a lonely and sometimes acrimonious battle with mainstream social scientists to keep the Darwinian lamp lit' (Scarr 1993: 1349). Unfortunately, 'Political correctness, rooted in fear of unpalatable results, seeks to stifle legitimate inquiry and intellectual discourse'.

Another indication of this modern imperative is the way that the treatment of the 'biology' or 'evolutionary background' of cognitive abilities is now mandatory in psychology texts and courses. The British Psychological Society, in a consultative report of a working party examining the undergraduate curriculum (1991: 5), admonishes that

> the biological approach . . . provides a range of important metaphors – about evolution, development, adaptation and change – which are fundamental to discussion to a wide range of issues in psychology. It would now be exceedingly unprofessional of teachers if they were not to include an informed discussion of the biological basis of topics as an integral part of the course.
>
> (British Psychological Society 1991: 5)

To their credit, though, the authors warn us of the 'hand-waving about the biological bases of our phenomena' in the past, and that these were 'often highly reductionist accounts'.

There seems, in fact, little to worry about from either point of view. Even a cursory survey of modern theory of cognitive ability indicates the prominence of the basic Platonic notions. In a wide range of theories, psychologists readily fall back on to forces (more or less mysteriously) 'innate' as the basis of cognitive abilities and their development, not just in the sense of hypothetical constructs awaiting empirical demonstration, but of ultimate formative power. These forces are inherited 'in' our genes at conception, and subsequently just grow or mature. As Light and Perret-Clermont (1991: 136) explain, 'Within academic developmental psychology the dominant view of cognitive development has always had a more or less maturationist flavour. . . . The basic ingredients of development . . . are given at birth. Development proceeds step by step, and involves the integration of the child's experiences into cognitive functioning'. And the metaphors of maturation and unfolding that bear it are frequently reinforced in the textbooks, as in common conversation: 'From conception to death the drama unfolds, revealing an ever more complex being in the making' says Rice (1992: vii). So it may be worth providing a few illustrations of that current dominance.

It is well known how the recent revival of Platonic nativism arose from Noam Chomsky's theory of language (Chomsky 1957a, 1957b). Pointing out that most of what a child hears in everyday language experience is highly diverse, faulty, and piecemeal; that a child actually produces numerous novel sentences that he or she has not heard before; and thus utterly rejecting the idea that the child learns grammar by imitating sentences heard, Chomsky returned to the idea that all the child's abilities in grammar must be inborn, in the form of a set of innate rules, fashioned by natural selection and coded in our genes.

This idea was soon taken up and applied to cognition generally – by Chomsky himself, as well as many others. By the 1970s and 1980s the computer became a useful metaphor for innate knowledge and reasoning processes, and the basis of an assertive 'Cognitive Theory' or 'Cognitive Science' (Casey and Moran 1989). Cognitive development could now be viewed merely as the 'maturation' or 'growth' of cognitive rules or programmes, without any of their essential structural form being learned or otherwise influenced by experience.

> My own suspicion is that a central part of what we call 'learning' is actually better understood as the growth of cognitive structures along an internally directed course under the triggering and partially shaping effect of the environment. . . . Our biological endowment determines both the scope and limit of physical growth. . . . Innate factors permit the organism to transcend experience, reaching a high level of complexity that does not reflect the limited and degenerate environment. . . . When we

> turn to the mind and its products, the situation is not qualitatively different from what we find in the case of the body. . . . We may usefully think of the language faculty, the number faculty, and others as 'mental organs'.
>
> (Chomsky 1980: 33, 39)

According to this view, then, the richness of our abilities is built up in genetic codes through natural selection. In views such as Chomsky's, these are present in all of us to the same degree; in others they vary across individuals. Either way, over the last twenty years or so this has, once again, become the dominant view of cognitive abilities and their development. Let us now try to get some idea of the enormous contemporary sweep of this doctrine.

Like Chomsky, Jerry Fodor (e.g. 1983) has proposed a system of information-processing 'modules' in the mind, one for each kind of content such as language, number, music, and so on. These are essentially automatic 'input' systems which analyse sensory inputs and deliver them to a 'central executive' in the form required. 'A module is an informationally encapsulated computational system . . . largely innately specified . . . and characteristically associated with specific neuroanatomical mechanisms' (Fodor 1985: 3–4). Others (e.g. Sternberg 1984) have proposed that cognitive abilities can be broken down into separate, step-like, information-processing 'components' that vary in processing efficiency from individual to individual. Gardner (e.g. 1984) has proposed a system of 'multiple intelligences' – distinct cognitive systems pertaining to different domains of ability or competence, but varying in strength from person to person so that we each have a particular profile of innate talents. 'The plan for ultimate growth is certainly there in the genome; and, even given fairly wide latitude (or stressful circumstances) development is likely to proceed along well-canalised avenues' (Gardner 1984: 56–57).

As just mentioned, a major presupposition of both Chomsky and Fodor is that our cognitive abilities are *modular*: i.e. discrete mechanisms, restricted to specific domains (domain-specific), rather than a single, all-purpose, 'domain-general' process. This is currently a very popular position, especially in the field of cognitive development. Thus, humans are said to be born with certain 'skeletal sets of principles that support the search for, and the coherent uptake of data that nurture further knowledge acquisition in, the domain they define' (Gelman 1991: 311). The 'subjects' of the school curriculum readily spring to mind as such 'domains', each said to reflect a different, genetically determined, cognitive architecture (see also Gardner's list of 'intelligences', mentioned above). Although within these domains development is 'constrained' by genetic codes, this is in different ways in different children because they will vary genetically. Thus, children may progress in language or drawing while being backward in mathematics or science. Indeed, the existence of children particularly advanced or 'gifted' in a specific domain is often taken as strong evidence that development is domain-specific and thus modular (Hirschfeld and Gelman 1994).

The current wave of theorising in this vein has invoked support from three or four sources. The first is the Platonic argument from 'the poverty of the stimulus': 'we find structures of considerable intricacy, developing quite uniformly, far transcending the limited environmental factors that trigger and partially shape their growth' (Chomsky 1980: 39). A second is the availability of useful metaphors in the apparent 'unfolding' of physical characters in humans. Indeed, from the formation of the embryo in the womb, to the growth spurt at puberty, it is easy to imagine a 'biological clock' operating. A third source of evidence is the vast range of complex reflexes and stereotyped, 'instinctive', behaviours (e.g. birdsong) in non-human animals. Reviews in Hirschfeld and Gelman (1994) marshal all of these, even though the direct applicability of such evidence to *cognitive* abilities in *humans* is hard to establish.

A fourth, and more recent, source of evidence seems to have clinched the issue for many cognitive and developmental psychologists. This is based on the assumption that, if our essential cognitive abilities reflect 'plans' in our genomes, then we might see signs of them at birth. A lot of effort in recent years has been expended in trying to demonstrate this in infants. Some genetically determined constraints must be present at birth, so the argument goes, otherwise the infant would not be able even to *begin* the complex work of cognition. For example, even the simple recognition of objects in space is more complex than it may seem because, in everyday experience, we rarely see a whole object 'at once': we may view it from an unusual angle, it may be largely obscured by something else, and some of its parts or features may be completely hidden from view. If infants are to learn anything about objects, some knowledge of 'objectness' must be innate, according to these recent theorists, and is evident in infants very soon after birth.

Much of the evidence sought to support this argument has come from 'recovery from habituation' or 'preferential looking' in infants. Infants, as young as a few hours or days in some studies, are asked to sit down in a small room with blank walls and an image of an object is projected on to a screen in front of them. They will tend to look for a period and then their attention appears to wane, so they look away – they are said to have 'habituated' to the stimulus. If the same stimulus is switched off for a few seconds and then switched on again, further looking is stimulated, but for progressively shorter times as the presentation is repeated. If, at some time following this experience, a *different* image is projected, the infant will look for a period as long as they did with the original – they are said to have dishabituated. Alternatively, infants may be presented with two or more pictures simultaneously, and the time they spend looking at each one – i.e. preferential looking – measured. Either way, the technique seems to present us with information about what infants can or cannot discriminate in their visual experience.

A classic study in this regard is that conducted by Kellman and Spelke (1983: 483) using a display like that in Figure 1.1. Four-month-old infants were habituated to a rod moving back and forth behind a block occluder. Following habituation to this arrangement, they were exposed, in turn, to the two stimuli shown at the bottom. The question was whether they would dishabituate more to the one on the left or to the one on the right. The latter, in fact, appeared to be preferred, even though the image is much closer to the original than the one on the left. In order to be preferred, so the argument goes, the infants must have been 'seeing' much more than just the *image* in the original; they must have been inferring a whole rod – and this depends on them having 'an inherent conception of what an object is'.

By using similar methods, various other 'innate abilities' or 'inherent conceptions' of young infants have been said to be 'revealed'. For example, infants have been said to show a preference for human faces over other objects (Slater 1989). Infants appear to have a preference for speech sounds over other acoustic stimuli (Mehler and Dupoux 1994). Human infants, it is claimed, are born with certain social predispositions or with 'innate interpersonal abilities' (Trevarthen 1983); 'humans are provided

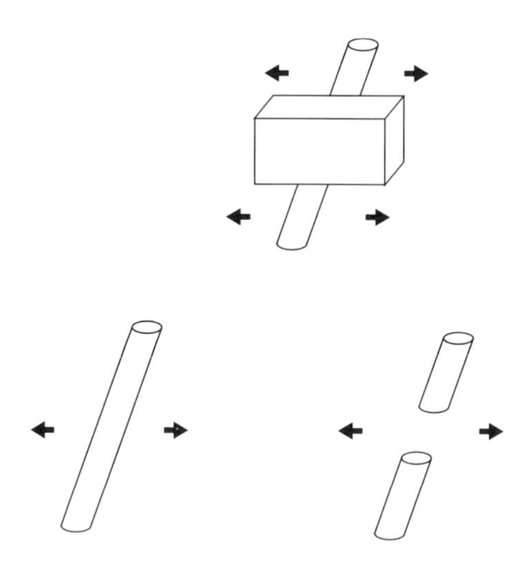

Figure 1.1 Infants are first exposed to the arrangement of rod and block (top), then their preference for each of the 'rods' at the bottom is assessed (redrawn from Kellman and Spelke 1983).

with an innate mechanism for social learning' (Meltzoff and Gopnik 1993: 352).

More recently, this idea has been expanded to include the innate cognitive structures that infants require to 'know' that other humans also have minds, and the 'intentionality' that goes with it. Infants are born, it is said, with an innate 'theory of mind' (e.g. Wellman 1990). The infant can perceive the 'outer-directedness' of people's attitudes because 'there are "hard-wired" mechanisms that guarantee this ability' (Hobson 1993: 121). In a recent summary, Spelke (1994: 433) argues that 'young infants appear to have systematic knowledge in four domains: physics, psychology, number, and geometry'. Or, as Mehler and Dupoux (1994: 98) put it, 'From space to objects . . . the baby seems to come equipped with a rich model of the world' which is 'the cognitive expression of the genetic heritage proper to the human species'.

A wave of such studies, data, and their interpretations has led, over the last decade or so, to renewed convictions about the gene-based cognitive preparedness of infants. Investigators admit that the procedures of data collection are extremely difficult, and in a typical investigation 50 per cent or more of infants will have to be rejected for 'fussing'. Moreover, the visual acuity of newborns is very poor; much development takes place between birth and four months; and no one really knows *what* newborns are looking at or preferring in such studies. None the less, the fact that even young infants *appear* to have such well-formed abilities has been very influential, radically switching theoretical perspectives about cognition and its development. Karmiloff-Smith (1992; 1993) suggests it has been quite decisive: 'The neonate and infancy data that are now accumulating serve to suggest that the nativists have won the battle in accounting for the *initial structure* of the human mind' (1993: 595, Karmiloff-Smith's emphasis).

A view of the gene

Of all the biological justifications for nativist assumptions in psychology, the strongest of all are thought to reside in the nature of the gene itself and the splendid descriptions of it that have emerged in the course of this century. In this section, therefore, I want to examine the commonly perceived nature of genes in their commonly perceived role as 'codes' for our cognitive abilities. Thus, I examine what is currently accepted among psychologists about the nature of the genes themselves; the nature and origins of genetic variations among individuals; the concepts of natural selection and adaptation; and how the information in the genes is seen to be 'expressed' in cognitive development. Then I shall show how these descriptions serve as presuppositions for current theories of innate structures (and also of individual differences). These will be the views generally presented in current theories and textbooks; but I will come to challenge all these presuppositions and arguments rather strongly in Chapters 2 and 3.

The simplest conception of genes is that of strings of chemicals whose composition 'codes' for our characters in some way, and which are replicated and passed across generations through the germ cells. The two most common metaphors of genes are 'beans in a bag' or 'beads on a string'. The first of these suggests the independent nature of each gene, and the randomness of the total collection in the 'bag'; the second suggests their arrangement as independent 'messages' on the chromosomes. This is the view commonly put implicitly or explicitly in popular accounts, and appears in psychology books and courses.

Let us first ask what exactly is the information that genes are said to carry from one generation to the next; what is it that is said to be 'coded' within them, and how?

The 'code' in the gene

Common observation of physical characters shows that offspring are very similar to their parents in all essential structures and functions, even though they have passed through a state when they were most *unlike* them. Long before Darwin's evolutionary theory, therefore (of which more below), it was realised that 'something' underlying these developed characters, and preserving that continuity across different forms, must be transmitted from parents to offspring. Apart from being likened to some sort of 'germ plasm', however, the nature of this 'something' was unknown until quite recently. Eventually, experiments by Gregor Mendel in the nineteenth century suggested their 'particulate' nature, and biologists at the turn of the century came to call them 'genes'.

It is now known that each cell of the body carries tens or hundreds of thousand of genes. The genes are contained within the nucleus of the cell, where they are arranged as long strings known as chromosomes (Figure 1.2). Different versions of each gene (called alleles) can also exist at the corresponding site or locus. And advanced (i.e. sexual) species carry *pairs* of such chromosomes (and thus two alleles, which may be identical or different, at the corresponding loci), one member of each pair derived from the mother the other from the father (Figure 1.2). So what is the information carried in each allele?

It was discovered only in the 1950s that each gene is a different version of a large chemical structure (molecule) known as DNA (Deoxyribonucleic Acid). DNA is made up of a few basic elements that can combine in different ways to form a unit – something like the way that a string of letters would form a word on a printing press (except that in DNA most strings are several thousand 'letters' long). In genes the 'letters' corre-

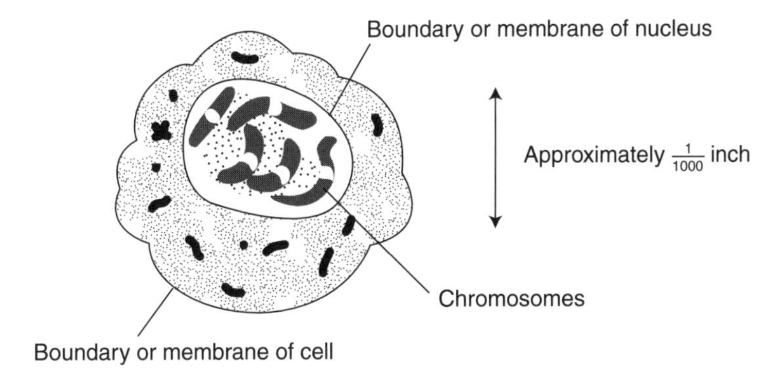

Figure 1.2 Genes are located on chromosomes in the nucleus of the cell where they usually exist as pairs.

spond structurally with an amino acid. Amino acids are the building blocks of proteins, so that each gene, in this chemical 'code', thus specifies the structure of a protein.

In the construction of a protein the gene functions like a chemical template, along which the specified amino acids are put together to form a string and, ultimately, the specified protein. The structure of the genetic medium is thus transferred to another, the protein molecule. This actually involves intermediaries, identified as different kinds of Ribonucleic Acid (RNA): messenger RNA (mRNA) and transfer RNA (tRNA). But the essential aspect of the process is that, in the translation, the thousands of configurations originally found in the genes are represented in the proteins and determine the latter's chemical properties.

Some of these proteins – the 'structural' proteins – form immediate components of the cells, such as membranes and transport vehicles. Others are the chemical catalysts called enzymes. These are the 'worker' molecules of the cell. By virtue of the specific structure 'picked up' from the gene, each enzyme brings about chemical reactions which produce the further constituents of the developing organism and catalyse the ten thousand or so chemical reactions necessary to the maintenance of life. It is in this sense that they are obeying the 'information' they have picked up from the genes. Figure 1.3 gives a simplified indication of this process.

At this point the gene is said to have been 'transcribed': its information has duly entered the developing system, and these corresponding protein structures will persist in *their* determination of the structures and functions of tissues and organs. So we get the standard impression of

genes ⟶ proteins ⟶ characters

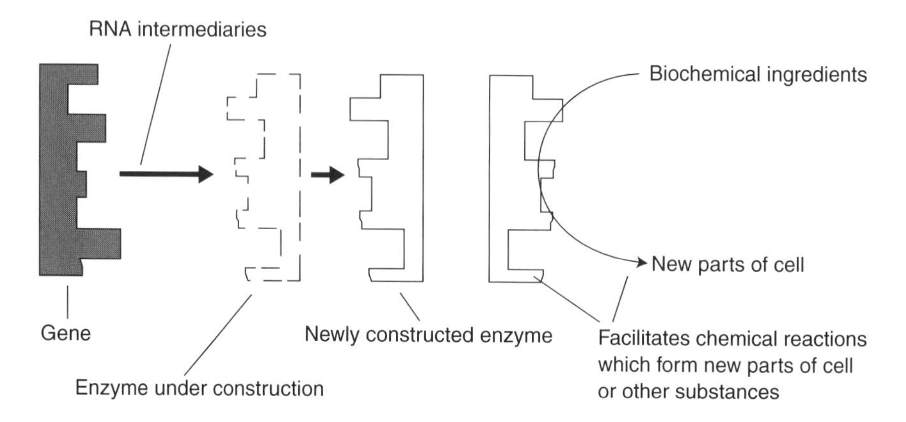

Figure 1.3 In the standard picture, genes act as a kind of template against which enzymes and other proteins can be constructed (one gene for each), although it is now recognised that the transcription process is more complex than this (see Chapter 2).

and the idea that genes serve as a kind of 'recipe' for making a human being, brains, cognitive abilities, and all. The last decade, indeed, has seen the birth of the Human Genome Project, with the aim of specifying the genetic 'recipe' of the typical human (see Lewontin 1994; and the next chapter).

The genes are also said to have the property of being 'self-replicating'. Under certain circumstances (see below, p. 18) their chemical structure is not copied into a protein but into another copy of itself. These copies then form the genes within the proliferating cells of the growing organism and, through the germ cells (ova and sperm) can be passed on to the next generation.

'Mendelian' versus 'polygenic' characters

Most of what we know about genes has come from studying variation in observable physical characters. From Mendel (1865, rediscovered in 1900) onwards, the 'laws of heredity' were elucidated by the study of characters whose variants fell into one or other of a few discrete classes (e.g. large/small; red/blue/yellow) as if due to the presence or absence of single genes. Mendel in fact chose clearly identifiable dichotomous classes among his peas, such as smooth or wrinkled seeds, colour of flowers, and dwarfism. These varieties are called different *phenotypes* (the different genes with which they are associated are called the *genotypes*). By cross-fertilising members of different classes and counting their representation in the progeny he arrived at his famous 3:1 ratio. That is, instead of a 'blending' of characters, as was at first expected, the original characters bred 'true', but by unequal proportions.

To understand this we have to recall that chromosomes occur in pairs (one from each parent) and that alleles at corresponding loci may or may not be identical. Where alleles or sets of alleles associated with a character vary, they are said to comprise different genotypes. It is usually the case that, where they are different, the expression of one allele eclipses the expression of the other; i.e. one allele is said to be 'dominant' over the other. If A is a dominant allele and a its recessive version, it follows that three combinations of alleles are possible as follows:

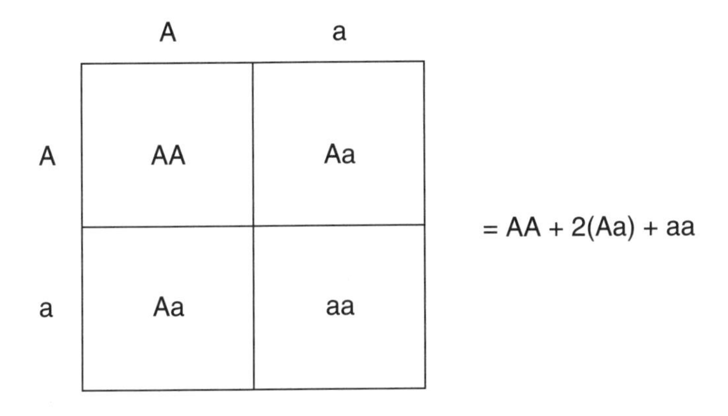

Since A is dominant, and Aa is thus indistinguishable from AA, the category associated with A will be expressed with three times the frequency of that corresponding with a. The Law of Segregation states that there are two hereditary 'elements' for each characteristic: these elements separate or segregate during transmission from parents to offspring, offspring receiving one of the two elements from each parent. The Law of Independent Assortment states that the hereditary elements for one characteristic assort independently of the elements for other characteristics. These principles have been enormously influential in models of genetic determination of development, creating quite distorted pictures of what are more evolved regulations.

Mendel's ratios clearly implied the particulate nature of the genes. But they also invited an emphasis on 'unit characters' and on genes as direct determiners of phenotypic characters, on a one-to-one basis. 'It has become evident that the individual is an entity made up of hereditary unit-characters. The aggregate of such units forms the individual. During the process of inheritance the units segregate, and by recombination form new individuals' (Herbert 1910: 10, quoted by Dobzhansky 1962: 32).

Although this view of genetic determination has persisted in some form in psychology today, it had to be relinquished in genetics when it was realised that not all characters are manifested as such discrete classes. Many characters of species, indeed, do not vary at all. Of those which vary, most, in fact, present a gradation of values in a group or population, as if depending on the expression of *large numbers* of underlying, but variable, genes, each having a small effect, as well as being influenced by the environment in which they are expressed. In humans, physical height is usually presented as the ideal model. This is now known as the *polygenic* mode of inheritance. Obviously, the identification of genes, of how they operate and how they vary, influence the structure of, and diversity in, the character, becomes infinitely more difficult with polygenic characters compared with 'Mendelian' characters. Attempts to overcome the problem have created considerable debate since the 1920s, and remains a problem today, as we shall see in Chapter 5. Throughout, though, the simple idea of direct expression from genes to characters, more or less helped or hindered by a 'good' or 'bad' environment, has prevailed.

Genetic variability

Offspring tend to be very similar to their parents in all essential respects. But they are different in many details, too. So how does this reproduction-plus-subtle difference come about?

An obvious possibility is that 'experience' or 'environment' has affected those details, but we have long known of another. Occasional 'mutations', or changes of chemical structure, can take place *within* each of the parents' genes, so creating new alleles. Most of these are said to occur spontaneously, though many have been traced to environmental agencies such as radioactivity and certain chemicals. If these occur in a parent's ovaries or testes where the eggs (ova) and sperm are produced, then they can be 'passed on' to the next generation.

The main consequence of this process of mutation is that genes or alleles at corresponding loci on pairs of chromosomes may be structurally different and thus 'encode' different information, as just mentioned. These differences become even more significant in the context of a second 'variation-producing' mechanism. In the production of eggs or sperm in males or females the members of the pairs of each chromosome go through a three-step process within the parent cells (Figure 1.4). First, copies of each member of each pair are made. Then the members of each pair separate. Just before they do, however, they may exchange genetic material by a process known as 'crossing-over'. Thus new 'admixtures' of genetic material arise. Finally, the separating members of the pair form

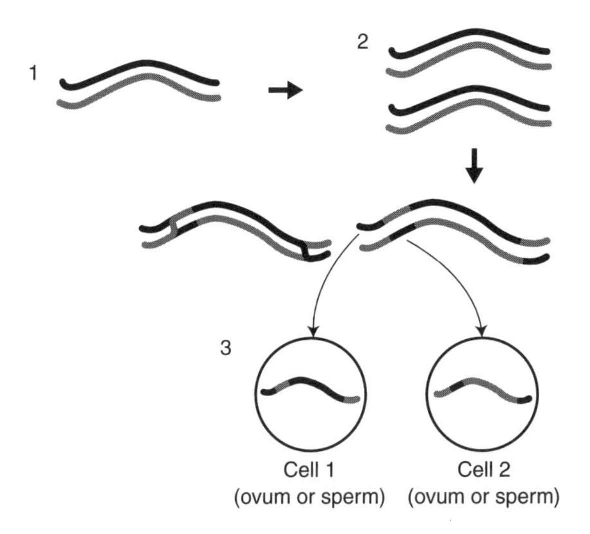

Figure 1.4 'Copying' of a pair of chromosomes and 'crossing-over' between copies before separation into cells (gametes) that will form eggs or sperm.

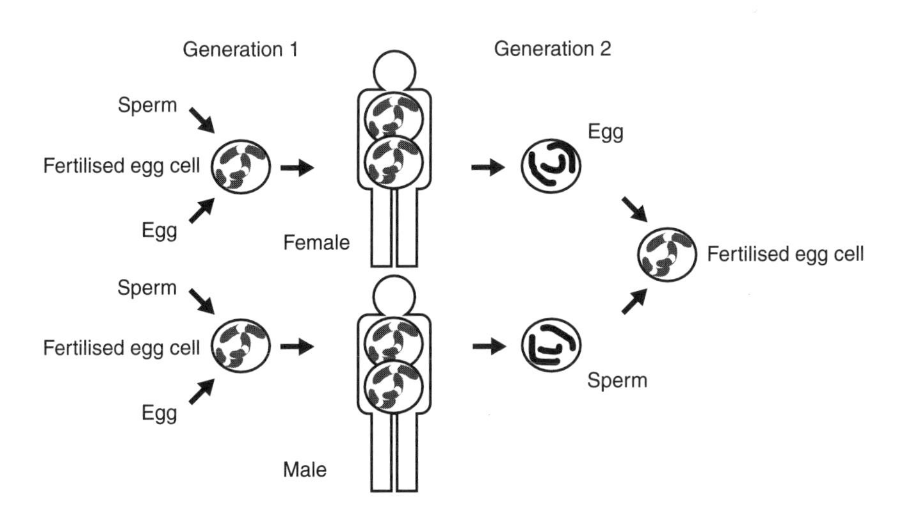

Figure 1.5 Diagram of the general process of replication of 'germ cells' (sperm and ova) each containing one set of chromosomes, and re-establishment of dual sets at conception. Humans have forty-six (twenty-three pairs) of chromosomes, containing about a hundred thousand genes in total, but only three pairs are shown here for simplicity.

parts of the cells that will form eggs or sperm (Figure 1.4). Pairs of entirely new permutations can then be 'made up' when eggs and sperm come together at conception (Figure 1.5).

The dramatic picture of the 'reshuffling' of genes across generations, and of how the set inherited by any individual will consist of a unique assortment, overall, has captured the imagination of popular biology and psychology alike. It is hardly surprising that it has been likened to a game of chance or a genetic lottery, generously or cruelly determining our fates, as the case may be. These processes are often referred to as the 'mechanisms of inheritance'. As described so far, the production of mutations, the replication of different genes, and their passage through and between families appear to be essentially random or chance processes. What a child gets, the 'potentials' thus inherited, and the characteristics he or she then develops, is thought of as 'chance' or 'luck'. All of these expressions have been used to explain structure and variation in cognitive ability, and now form part of the scientific understanding of the general public as well as most psychologists.

Natural selection and adaptation

A moment's reflection ought to show, though, that each individual human is more than a random 'bag' of genes: there is cohesion as well as diversity. Of course, we most notice variations such as eye colour, hair colour, height, and shapes of noses in children as they develop. What we notice much less is how closely alike they are in sharing hundreds of identical physiological processes, numbers of limbs, teeth, ribs and vertebrae, and, more importantly, the regulatory processes through which individuals are responsive to the environments they actually experience. So the general assumption of a kind of universal 'law' of genetic and phenotypic variation needs to be qualified.

How is it that we can be so alike in some characters yet different in others? This brings us to consider two concepts which dominate biological aspects of psychology: 'natural selection' and 'adaptation'. These concepts have been highly influential in the recent theories of cognitive ability that have sought foundations in evolutionary biology, so it is worth looking at them fairly closely.

Natural selection is at the heart of Charles Darwin's (1859) theory of evolution. Basically, it suggests that random genetic variations are produced, and 'passed on' to the next generation, by random mutations and recombinations as just discussed. These produce corresponding variations in each developing characteristic in the population – i.e. different phenotypes. But some of these phenotypes will be 'fitter' than others in the environments they encounter. For example, it might have been, at some stage in our past, that variation in genes resulted in variation in, say, manual dexterity, with some individuals more dexterous than others. Children receiving more favourable genes will grow up to be better tool-makers, gather more food, and produce more offspring that themselves survive and reproduce. Accordingly, a higher proportion of individuals in the next generation will have those very genes. And so the process goes on, with each successive generation tending to become more alike genetically, with respect to that characteristic.

Thus, the standard and widespread assumption of an unlimited and unconstrained genetic variation underlying all our characters, including cognitive ability, needs to be qualified: natural selection, in the traditional model, tends to 'weed out' genetic variation underlying characteristics important to survival (although in specific cases there may be 'balancing' or other forces which maintain some genetic variability, which I will not go into here). Meanwhile, of course, variation in characters not (currently) so crucial to survival is not selected and is 'allowed' to accumulate. Where such characters can be quantified (body size or human height are often taken as classic cases) the 'admixtures' of variable genes in variable environments often result in a 'normal distribution' of frequencies of values – average values being more frequent than more extreme values (Figure 1.6). In other words, non-selected characters tend to assume a *distribution* in a population. The coexistence of selected and non-selected attributes can explain why James may have his 'mother's' nose, and his 'father's' eyes (implying genetic variation between mother and father), yet share with both parents (as well as the vast majority of other humans) a vast number of common characteristics (and common genes).

It is clear that most of the prominent 'species-specific' traits have been subjected to natural selection, with consequential reduction in genetic and phenotypic variation, and numerous and complex deviations from this

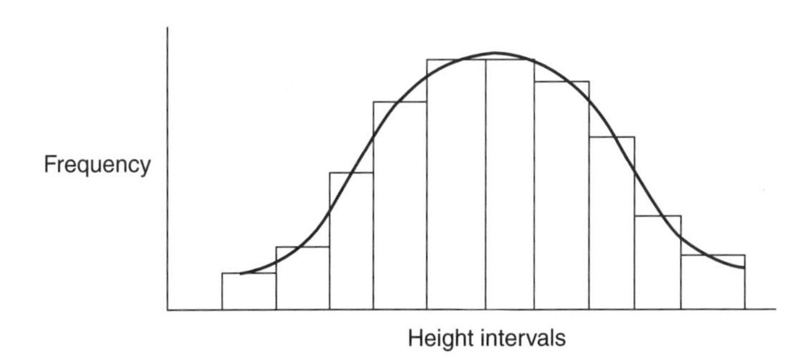

Figure 1.6 If height is measured across a population of individuals and the results are allocated to intervals (of, say, one inch) the frequencies in each category will collectively approximate a 'normal' or bell-shaped curve.

simple picture of the normal curve. The clearest examples are numbers of limbs, teeth, vertebrae, ribs, digits, and a vast range of common metabolic processes. In Chapter 5, when considering attempts to explain individual differences in cognitive ability to genetic differences, I will note how often this simple 'law' of natural selection is overlooked. We need to be cautious about views like that of Scarr (1993: 1335) who claims that 'All periods of the lifespan have been subject to selection and are genetically variable to some extent' – which may be a contradiction in terms. Compare this with the fact that, as Rendel puts it,

> Some characters are expressed to exactly the same degree in almost all members of a population. Uniformity of phenotype is most obvious when the character is countable, such as the number of fingers or toes in a vertebrate, or the number of bristles in the scutellum of an insect. While a population can also be quite uniform with respect to characters that are not countable . . . uniformity for uncountable characters is not so easily measured.
> (Rendel 1967: 42)

One implication of the principle of natural selection, then, is that traits which are evolutionarily important will tend to exhibit little if any genetic

Figure 1.7 Heads of some of the finches studied by Darwin, showing the correlation between beak structure and feeding habit (after Darwin 1839). (1) a large seed eater; (2) small seed eater; (3) small insect eater; (4) large insect eater.

variation; or, conversely, traits exhibiting appreciable genetic variation have probably not been evolutionarily important. The assumption that there is unbounded genetic variation underlying all characters has become taken for granted in nature–nurture debates, although biological reality is far from being quite as anarchic. This presents a paradox for behaviour geneticists like Scarr (1993), and we will be returning to it in Chapter 5.

The reason that some phenotypes are selected in preference to others is that they are 'fitter', by virtue of being better 'adapted' to a particular aspect of the environment in a life-promoting way. Thus, increased manual dexterity in conditions where food is relatively inaccessible or in difficult locations may well be better adapted. The most famous examples of adaptation are the finches described by Darwin, whose beaks are superbly designed for the food they eat (Figure 1.7). Hence we arrive at the modern picture of natural selection 'acting' on the more adapted characters, and thus the genes associated with them, so that more of those individuals who possess those genes will survive, with the latter achieving a bigger representation in subsequent generations, and so on.

Evolutionary underpinning of nativist psychology

Evolution-as-adaptation has become a powerful notion in discussions about mental ability. Once taken on board, adaptationist assumptions can be used to 'explain' almost any cognitive form and cognitive variation, even in the absence of any evidence for it.

Here I can only give brief illustration of the 'evolutionary reasoning' flowing from these assumptions in modern psychology. A fairly typical version of a 'selectionist' and 'adaptationist' perspective on the *forms* of human cognition is put by Cosmides and Tooby (1994). They argue (1994: 85) that the abilities we have now are 'evolved adaptations, produced by the evolutionary process acting on our hunter-gatherer ancestors' over a million years ago. These adaptations are determined by 'the situations our ancestors encountered as Pleistocene hunter-gatherers', thus defining 'the array of adaptive problems our cognitive mechanisms were designed to solve'. One consequence of this is that 'there is no warrant for thinking that selection would have favoured cognitive mechanisms that are well-engineered for solving classes of problems beyond those'; we cannot 'solve all problems under all circumstances . . . .because our species did not encounter all problems under all circumstances' (1994: 87). Thus, thanks to the kinds of genes selected to furnish the kinds of abilities needed to solve the kinds of problems encountered in the past, the kinds of problems we are able to solve today are strictly limited.

It is only a short logical step from such a view to that of the mind as a set of cognitive 'organs', each adapted to particular kinds of (durable) problems in our evolutionary environments. This is the basis of the recent enthusiasm for the 'domain-specificity', or 'modular' theories, described earlier. As Hirschfeld and Gelman (1994: 118) argue, 'a domain-specific mental architecture is an inevitable consequence of human biological history and neurostructure'; or, as Mehler and Dupoux (1994: ix) put it, 'Like other animals, humans are able to adapt . . . only in certain areas and within certain limits. . . . While these [cognitive] skills can evolve, they can do so only within the confines of a narrow genetic envelope, which imparts to members of the species a fixed core of aptitudes'. Even for our most varied cognitions around objects, innate knowledge 'has been internalised through evolution so that it is now 'pre-wired' in individuals and governs how they apprehend objects in space' (Gardner 1984: 129).

Within this simple picture, the explanation of the origins of all our cognitive abilities becomes reduced. By direct analogy with Darwin's finches we have fixed cognitive 'beaks' for specific cognitive 'challenges' presented in the environment. Reasoning processes, knowledge structures, remembering, problem-solving, our joint actions with others, are analogous to physical organs and functions in respect of their origins and current forms. More sophisticated versions may refer to 'constraints' or 'epigenetic' processes (see Chapter 3), but these are never characterised and the basic cognitive form remains specified in the genes (e.g. Karmiloff-Smith 1992; Gelman 1991).

The two-layer theory

Another assumption involved in this evolutionary genetic determinism is that organisms, including humans, basically consist of two levels of regulation. The first is the all-determining genetic level of regulation. The second is the set of characters constructed from it, including cognitive abilities. These characters are a kind of outer 'shell' expressing the potential of the genes within. Genes are seen as potential: abilities as the outcome – or 'realized potential' as Kimble (1993) starkly puts it.

This assumed relationship between the 'inside' and the 'outside' is implicit in many statements.

> There are hereditarily determined, biologically determined, characteristics of intelligence that will provide a measure of potential if we can measure them.
>
> (Detterman 1993: 40)

> In a sense all behaviour must be coded for by genes; reduced to its simplest form behaviour is nothing more than a series of nerve impulses and muscle contractions and the protein structure of nerve and muscle is coded for by genetic instructions.
>
> (Krebs and Davies 1981: 12, quoted by Bateson 1988: 200)

This 'two-layer' assumption, then, permits us to view cognitive abilities exactly like any other character, as far as their structural determination and individual variation are concerned. This is the case whether or not they have been subjected to natural selection, and whether or not this has resulted in any greater complexity: there is a singular biological process operating in the expression of the genotype into the phenotype. 'The ways in which genes shape an individual's perceptual and cognitive capabilities influence the propagation of those genes in the species ecological niche just as much as the way in which those genes shape the individual's size, shape and coloration' (Shepard 1994: 2).

Thus, it is assumed that genes 'for' cognition, and genes 'for' height, act to produce a bell-shaped distribution of phenotypes for one every bit as much as the other (Herrnstein and Murray 1994). It is generally assumed that differences in cognitive ability, say, arise by much the same mechanisms as differences in height. The genes predetermine the limits of the characters' development, and individual differences in it, with the environment 'nourishing', but having little formative influence: 'Feeding a well-nourished but short child more and more will not give him the stature of a basket ball player. Feeding a below average intellect more and more information will not make her brilliant' (Scarr 1992: 16). This reduction has encouraged the use of highly simplistic models of gene–environment causation in which cognitive abilities are treated merely as the 'summary' of good or bad genes and 'good' or 'bad' environments (Figure 1.8; see further in Chapter 5).

Another consequence of the two-layer assumption is the view that there is little true development – what we see on the outside is simply an 'automatic' expression of what is on the inside, however complicated that may be:

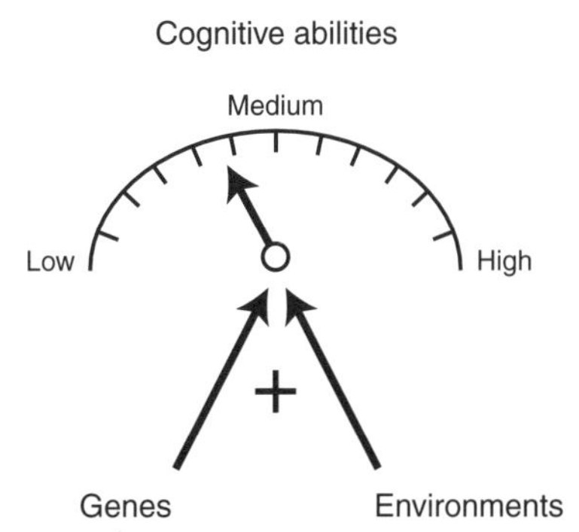

Figure 1.8 Cognitive abilities reduced to the addition of 'genetic' and 'environmental' forces in a simple quantitative character, prominent in nature–nurture debates in psychology.

> The regulation of developmental processes over the lifespan is accomplished through the gene-encoded production of hundreds of thousands of enzymes. . . . Development proceeds on a gene-regulated pathway of enzymatic activity.
>
> (Scarr and Carter-Saltzman 1982: 807)

In this view cognitive abilities are no more 'developmentable' than an arm or a leg. Indeed it has been claimed that we could, in theory, 'compute the adult organism from the genetic information in the egg' (Wolpert and Lewis 1975; quoted by Goodwin 1985: 49). This is sometimes called the 'maturational' view, but we can also call it the 'assembly-line' view. Figure 1.9 shows the 'assembly-line' model of predetermined development from genes to cognitive structures, implicit in such views. The general idea is that the structures of the initial parts are specified on the basis of the genetic codes. These then amalgamate to form 'bigger' structures, such as patterns of nerve–cell connections, or 'architectures', in the brain, which, in turn, determine their functions. Throughout, a correlation between organic structure, cognitive function, and the essential formative power in the genes is maintained.

Note that maturationist theorists will mention the availability in the course of development of appropriate environmental factors, with which the genetic instructions 'epigenetically' interact. But these are usually only seen as attenuating or enhancing maturation, without materially altering the fundamental structure or limit of the end-point. Of course, in the case of cognition and cognitive abilities, the whole assembly process is an assumption, of which we have no empirical demonstration.

A popular version of the 'assembly-line' model of development has been that of a 'genetic programme'. This, too, has been applied equally and uncritically across both physical and cognitive characters. In psychol-

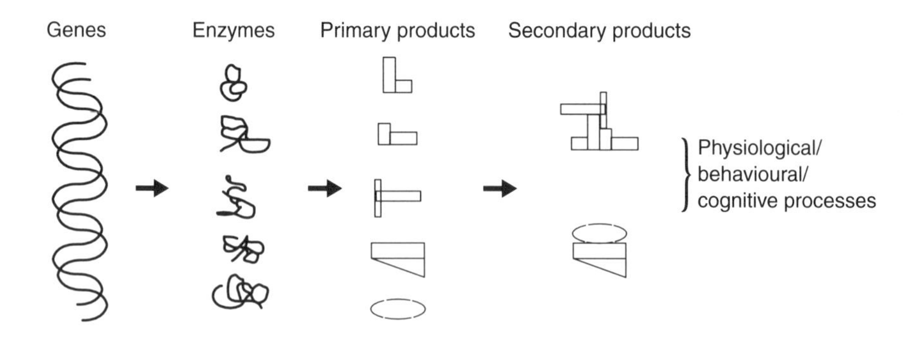

Figure 1.9 'Assembly' model of cognitive development.

ogy the idea stems largely from the work of Arnold Gesell (e.g. Gesell and Ames 1942), and his observations of the early motor development of children. He concluded that the orderliness and predictability of the developmental sequence could only be accounted for by an internal 'programme' of development situated in the genes.

> Gesell believed . . . the child's development is directed from within, by the action of the genes. Gesell called this process maturation. . . . An outstanding feature of maturational development is that it always unfolds in fixed sequences. . . . This order, which is directed by the genetic blueprint, is never violated.
>
> (Cain 1992)

A similar idea of genetically determined maturation of abilities underlies the construction of 'intelligence' and other standardised ability tests (see Chapter 4). Again, a close analogy is drawn between physical and cognitive development: as Keil (1988: 95) put it, innate 'constraints are either present throughout development or emerge at certain points in a maturational fashion relatively independent of experience, just as beards emerge in men'.

A good example of the way evolutionary stories and adaptationist assumptions are used to 'explain' cognitive functions concerns the differences in performance on certain spatial reasoning tasks on which males consistently tend to do better than females. It is argued that this reflects sex-linked genetic differences arising by selection, in males, for better navigation, hunting, and gathering skills among our ancestors a million or more years ago (Silverman and Eals 1992). It has been suggested that a similar evolutionary story explains differences between males and females in mathematical achievement in schools (Benbow and Lubinski 1993; see Geary 1996, and peer-commentary therein). Again, note the 'adaptationist assumption' that all selected characters, like the beaks of Darwin's finches, are stable fixtures, selected to 'fit' equally stable aspects of the environment.

Rather more extreme has been the conclusion that all human behaviour and human culture – the structure of institutions, economies, and relationships – are the products of what has been coded in the genes through the hand of natural selection. By a reverse current of reasoning (perhaps because they have seen the field wide open) many biologists have taken it upon themselves to explain to psychologists, as well as the general public, what human cognitive abilities and other aspects of behaviour are *really* like. Although almost entirely based on extrapolations from non-human animals such as ants, bees, and fish, this theoretical current has blossomed into a new discipline called sociobiology (e.g. Wilson 1975). According to a typical sociobiological thesis, variable perceptions and cognitions are related to variable genotypes, and have evolved within a culture because of a need for divisions of labour and expertise; these divisions may extend across cultures, too (Lumsden and Wilson 1981). In this 'total synthesis' psychology itself, along with other human sciences, is swallowed up, so that Rose, Kamin, and Lewontin (1984: 239) can speak of 'The intellectual imperialism of a new discipline that threatens to engulf all other intellectual domains'. Though attacked by large numbers of critics (e.g. Rose, Kamin, and Lewontin 1984; Kitcher 1985; Beckwith 1987), this thesis appears to have further inspired 'evolutionary' accounts of ability within psychology.

Even more dangerous, of course, is the way that simple assumptions about evolution and cognitive abilities have repeatedly created a scientific racism in which evolutionary or adaptive stories are imaginatively constructed to 'explain' a ladder of racial superiority. There have been many such accounts, usually involving IQ as a supposed measure of cognitive ability, and I will be returning to them in Chapters 4 and 5.

Summary of assumptions

When all these assumptions are put together, the story constructed offers a superficially compelling model of the origins of cognitive ability and individual differences in it. It is widely accepted by cognitive and developmental psychologists, and resonates closely with impressions held by the general public and with actual social structures. These are the assumptions dealt with so far.

1 The assumption that cognitive (like other) characters can be 'innate', in the sense of preformed or predetermined by some sort of singular active agent.
2 The assumption that genes act as 'codes' for characters.
3 The assumption that the genes we possess have been selected from a pool of antecedents because they determine physical and cognitive structures which 'fit' particular aspects of the environment: this is the adaptationist assumption, and it is assumed that it applies to physical and cognitive characters in exactly the same way.
4 The assumption that, regardless of such selection, the genes underlying physical and cognitive characters will always continue to have functionally significant variation across individuals in a population.
5 The assumption that the manifestation of a phenotype is simply a faithful expression of the code in the genes – i.e. there is little if any true development, only a maturational inevitability, or growth, of what is 'already there'.
6 Because genes are conceived as solitary powers of different 'strength', and because of the way they are conceived to vary across individuals, this manifestation results in a corresponding range of 'strengths' among phenotypes, the distribution of which approximates a normal curve.

In spite of their overwhelming dominance in current views, I want to show that all of these assumptions are highly questionable or incorrect, at least for any general model of causation of complex characters like cognitive ability. Most of this critique is offered in Chapters 2 and 3 (where I also offer alternative views). In Chapter 4, I offer a critique of the notion of IQ, an instrument deeply rooted in the same assumptions (with the addition of others, which I also criticise). In Chapter 5, I evaluate some of the evidence thought to demonstrate that individual differences in cognitive ability are, at least in part, direct reflections of genetic differences, and show how it relies on the veracity of these and additional assumptions. In Chapter 6, I consider how the same assumptions have played a critical part in education, and put forward some alternative views.

Of course I hear the reader looking for traditional balance and saying, 'OK, but what about assumptions about the environment – why aren't they up for scrutiny?' The reasons for this apparent neglect are simple. First, the view detailed above, or something close to it, is the overwhelmingly accepted one in modern psychology. The second is that, because of this, study of the 'environment' relevant to cognitive development has been almost entirely neglected, except as highly general and vague 'factors' (social class, family size, etc.) working as independent forces, or else as 'anonymous components of variance' (Plomin and Thompson 1993: 80). The unfortunate tendency has been to view environments as sets of (independent) 'good' or 'bad' factors that merely assist or attenuate gene expressions, rather as we view the role of nutrition in height. *The uncomfortable truth is that, beyond these vague hunches, there are few assumptions about the nature of the environment around.* I will have much more to say about this unfortunate state of affairs, and about what counts as the environment, in human cognitive development, in Chapters 3 to 6.

All this is to follow. But I think it is first important to consider some other implications arising from the story described so far. This is because I think they add further urgency to the critique, and its alternatives, which follow in Chapters 2 to 6. Such consequences of the assumptions described so far is what the rest of this chapter is briefly about.

Explaining cognition away

Much has been imagined, therefore, on the basis of a few simple assumptions about evolution and the nature of genes. The most important major aspect of this imaginary process has consisted of a general tendency to attribute a kind of predestiny to the genes. This is known as genetic determinism. The view of the nature of the genes' involvement in development is given away by the terms and metaphors used. These include 'determination', 'control', 'programmes', 'directed by', 'instructions', 'regulation', 'constraints', and so on. All of these give the impression of some fundamental 'thinking' power, the creative force of development, intervening at every stage, fundamentally determining individual differences, forever pervading our lives, and against which we can do nought but submit ourselves fatalistically. As already mentioned, Piaget (e.g. 1929) described such causal attribution as 'intentional causality', in which an agent is seen as a purposeful initiator of events, and which is very common in young children. This is a view of 'the gene as animistic metaphor, a kind of ghost in the biological machine' (Oyama 1985: 7) or 'a mindlike force to fashion matter into a functioning animal-machine' (Oyama 1985: 29).

As already mentioned, though, this view has flourished in the absence of any direct empirical support for it. Indeed, a conspicuous feature of nativist theorising around cognitive abilities has always been its theoretical *vacuity*: a tendency to 'explain' cognitive regulations by, as it were, simply consigning them to unspecified biological forces. As Johnston (1994: 721) points out, 'nativist accounts are usually silent on the question of how . . . (innate) elements come into being. The strategy has been criticised by several generations of developmental theorists . . . on the grounds that invoking the concept of innateness amounts to abdicating responsibility for explaining development'. And as Cain (1992: 53) notes with reference to Gesell's maturationism, 'The precise mechanism by which genes work, in Gesell's day as today, is still mysterious'.

The *content* of what is innate, and what develops, thus remain shrouded in mystery. For example, innate 'constraints', 'attentional biases', 'potentials', 'predispositions', and 'rules' are commonly envisaged (see contributions in Carey and Gelman 1991, and in Hirschfeld and Gelman 1994), without specification of what these are. Sperber (1994: 51) says that 'All that the internal structure provides is . . . a mode of construal, a disposition to organise information in a certain manner and to perform computations of a certain form'. Karmiloff-Smith (1992: 595) argues that the human cognitive apparatus has 'some innately specified information that allows it to attend to persons, space, cause and effect relations, number, language, and so forth'; and suggests that humans have 'a number of innately-specified processes which enable self-description and self-organization', together with 'mechanisms for inferential processes, for deductive reasoning and for hypothesis testing'.

All of these are, of course, very stimulating, complex ideas. But many authors have pointed to their vagueness. What manner? What form? What mechanisms? These are questions we are entitled to ask. What would we think of physiologists who, after decades of research, came up with the conclusion that digestion (innately) processes food 'in a certain manner', or that the kidneys 'perform filterings of a certain (innate) form'?

This is merely the enumeration of functional ends, not a description of functional *means*. Indeed, in area after area, wherever a phenomenon has been established (theory of mind, face processing, analogical reasoning, etc., etc.) instead of psychological theory, psychologists 'explain' it by telling us that it is a biological 'given'. This strategy is seen in another descriptive tool frequently used by cognitive nativists. This is the flow-chart of 'information' passing from module to module, or, more recently, 'architectures', which are assumed to do the essential work by virtue of simply being part of our genetic endowment. The almost universal strategy in cognitive theory has been to devise computational models to emulate outcomes in highly restricted tasks such as object recognition, without explaining how they are implemented in real brains and minds. Wherever a cognitive function is identified, there is a tendency, in other words, to simply invent a genetically determined 'module' or 'architecture' to 'explain' it. As Dennett (1978: 59) put it, 'each box specifies a function without saying how it is to be accomplished (one says, in effect: put a little man in there to do the job)'.

As already hinted, theorists often seem to be trying to have it both ways. For example, Karmiloff-Smith (1992) argues in favour of early genetic constraints, and subsequent 'epigenetic' cognitive development within those constraints. Some describe the biological and social as part of a harmonious stream, and thus view 'biological processes of growth and development as being continuous with psychological development, and the human mind as a product of a developmental unfolding of biological potential in a given sociocultural context' (Plunkett and Sinha 1992: 212). Keil (1988: 92) envisages a 'conceptual core or core competency that is defined by sets of rigorous constraints and which is supplemented by a periphery that is governed much more loosely'.

This may sound very reasonable: but, in the context of vast diversity in human cognitive functioning, the limitations of such formulations need to be stressed. Indeed, they appear to be based on rather loose metaphoric reasoning. For example, Keil (1988: 92) argues that 'dramatic shifts can occur in the periphery provided they honour the general boundaries of the skeleton. Because the core is only a skeletal framework, it does not completely constrain the structures within it'. But real life structures, anatomies, and architectures that are associated with an underlying skeleton remain closely tied to and determined by it. Dramatic changes in animal anatomies do not occur without changes in the skeleton; and we can hardly argue that dramatic changes in the shape of a building, say, can change without a change in its foundations. Yet such dramatic changeability is precisely what most characterises human cognition.

In other words, theorists are extremely vague about these genetic constraints, and there appear to be no compelling empirical grounds for suggesting that they exist for cognitive abilities at all. The main argument, as already mentioned, stems from studies suggesting 'early competency'.

Indeed, modern nativism has been given a strong boost by research in infants and newborns, suggesting that they are 'born' with some well-structured perceptual, cognitive, and social abilities. For example, it is suggested that, because four-month-old infants respond to 'surprising' events (such as the discontinuous stick shown in Figure 1.1 (p. 12), or a ball apparently having passed through a solid barrier) much as adults do, then they must have adult-like conceptions of object properties as part of their genetic endowment (e.g. Spelke *et al.* 1992). I also mentioned earlier the enormous methodological difficulties surrounding such work. When over 50 per cent of subjects have to be rejected for fussing; when the experimental procedures are extremely complex; and when interpretations are made on the basis of, usually, marginal differences in a few seconds of looking time at one picture rather than another, we obviously need to be cautious about what we make of the results.

I cannot engage in a detailed critique of all experiments here, although it is true that results have often been replicated and may well point to something very interesting. What is *most* interesting is the swiftness with which they have been taken to indicate 'innate' abilities in the genes – almost as if there is a deeper 'need' to do so. Karmiloff-Smith (1992; 1993) describes how such studies have 'won the battle' in accounting for the initial structures of the mind. But Spelke *et al.* explain where this victory leads us:

> It rejects several classes of explanation for the foundations of cognition, but it offers no explanation in their place. The apparent emptiness and arbitrariness of nativist proposals in psychology are characteristic, and they lead characteristically to discontent. . . . If cognition is part of humans' psychological beginnings . . . psychologists cannot contribute to the explanation of its origins. That explanatory task falls entirely to other disciplines.
>
> (Spelke *et al.* 1992: 629)

In citing such remarks, Thelen and Smith (1994: 33) describe how they make transparent a self-defeating dualism:

> We believe that developmental psychology is in grave danger if it gives itself the right to say 'innate' whenever there is as yet no developmental theory and then frees itself from saying what innate 'means' by saying that solving that problem is someone else's job.

It may be, as Thelen and Smith suggest, that a great deal of development has already taken place in four-month-old infants, who, by then, have had a tremendous amount of experience in the physical, as well as the social, world. In addition, what infants may be preferring in their preferred looking may be complex aspects of stimuli hitherto unrecognised. Elsewhere I have argued (Richardson 1992; Richardson and Webster 1996; see Chapter 3) that all a cognitive system needs to be sensitive to, in order to be maximally adaptable, is the nested covariation structures with which the world of objects and events is replete (see Chapter 3). If this is the case, then what infants need to be born with is not preformed cognitive structures, but acute sensitivity to in-depth covariations in the environments they actually experience. The kinds of stimuli which infants appear to prefer – faces and other social events; speech sounds compared with other sounds; the 'broken' as opposed to the continuous stick in Figure 1.1 – all exhibit more complex covariation structures. The point is, of course, that the flight into innate structures has led us to neglect the possibility of alternative, *developmental*, stories.

Theoretical backwardness

One of the most unfortunate consequences of this reliance on the origins of cognition, and its diversity, in, as it were, 'another place' – one into which psychologists cannot pry – is the theoretical fragmentation that prevails around cognition and its development. The truth seems to be, alas, that the subject retains many of its secrets, and any general cognisance about cognition still seems remote.

As an indication of this scientific backwardness, let us look at what the *Penguin Dictionary of Psychology* has to say about cognitive ability. Here we find 'Ability' defined as 'The qualities, power, competence, faculties . . . talents etc. that enable one to perform a particular feat at a specific time'. But this does not tell us what these 'powers', 'faculties', 'talents', etc. are, nor how they 'perform' – at least no more than a member of the general public would. Perhaps these will be specified under '*cognitive* ability'. The nearest referent is 'Cognition' which is described thus:

> A broad (almost unspecifiably so) term which has been traditionally used to refer to such activities as thinking, conceiving, reasoning etc. Most psychologists have used it to refer to any class of mental 'behaviours' (using the term very loosely) where the underlying characteristics are of an abstract nature and involve symbolizing, insight, expectancy, complex rule use, problem-solving . . . and so forth.

Obviously, terms like 'unspecifiable' and 'loose' hardly suggest definition and precision. But a more comprehensive survey of scientific research and theory about cognitive ability doesn't either. Many authors have expressed their disappointment when they have come to look at the psychology of cognitive abilities. For example, Scholnick (1994: 728) argues that there is no common view of contents of cognition, or of cognitive development, and that 'The current cognitive map consists of fractionated territories, each inhabited by processors who use specialized and untranslatable languages'. Beilin (1987: 37) says that 'one sees a bewildering variety of developmental theories. . . . To some observers, theoretical discourse in developmental psychology has the sound of the Tower of Babel: many tongues and little communication among the speakers'. And Hirst and Manier (1995: 89) argue that cognitive psychologists have led us into a 'stuffy windowless room, shut off from the "real world" – a reclusive chamber of our own making', from which we fail to answer even simple everyday questions like why we have difficulty remembering names, or how we solve simple problems.

In the absence of coherent theory, psychologists, again, evoke cognitive spirits. The (unspecified) formative power attributed to the genes is one such spirit. Another is seen in the way that psychologists frequently attribute cognitive diversity to an underlying 'general ability' (with such diversity determined by genetic diversity). As I shall show in Chapter 5, psychologists have never been able to describe this 'general ability' in any way that they can agree about, and there is little objective evidence for it: yet they cling to the idea with remarkable tenacity.

Constraints on intervention

There is another, very important, consequence of the assumptions listed above: they automatically foster a fatalistic view of human development in cognitive ability as in height or other physical attributes. Just as they are vague about means and contents of cognitive development, nativists are nebulous about targets and possibilities for intervention in development, such as education.

The strongest principle proposed is that of innate cognitive potentials which 'will out', given the freedom and suitable exercise to do so. In the writings of the great progressive educators, such as Pestalozzi and A.S. Neil, it consists of the idea that children must have the freedom of exercise and choice over activity which best suits their innate potentials. Froebel, for instance,

> agreed with Pestalozzi that children are born in possession of specific mental faculties, but he asserted that growth, power and self-fulfillment develop from the child's inner impulses, through spontaneous activities. . . . As plants grow from within, aided by nature, so the child will unfold and grow if aided by parents and teachers to express his instincts and utilise his native powers.
>
> (Smith 1979: 165)

As Maria Montessori put it: 'The fundamental principle of scientific pedagogy must indeed be the liberty of the pupil; such liberty as shall permit a development of individual, spontaneous manifestations of the child's nature' (quoted in Isaacs 1966: 156).

To Scarr (1992) the message of the model is that

> it is not easy to intervene deliberately in children's lives to change their development, unless their environments are outside the normal species range . . . for children whose development is on a predictable but undesirable trajectory and whose parents are supporting a supportive environment, interventions have only temporary and limited effects.
>
> (Scarr 1992: 16)

In other words, developing children might be able to cope with a wide range of environments; but they cannot cope with their genetic lot which really determines their fates.

These are, of course, principles of radical *non*-intervention, based on a belief that development is already 'written' in the child's 'nature' and, therefore, that we do not really *need* to know much more about it. A similar principle that (innate) *differences* in cognitive abilities 'will out' has, of course, led to justification for educational selection and the tacit belief that we can do little about a predetermined range of inherent potentials.

Nativism, superstition and ideology

Fatalism and mysticism are, of course, often found together in the intuitions of casual observation, but it is the task of scientific research to overcome them. So we seriously need to ask why the attribution of 'intentional causation' to genes remains so strong in modern psychology. Oyama (1985: 26) sees such superstitious attributions as part of a Western cultural tradition: 'Just as traditional thought placed biological forms in the mind of God, so modern thought finds ways of endowing the genes with ultimate formative power'.

On a wider stage, though, it can be argued that the gene doctrine has repeatedly re-emerged *precisely* to reinforce a Platonic 'social-rationalism', in which the state is run on genetically determined principles, and under which everyone accepts their predetermined place. As Rose, Kamin, and Lewontin (1984: 9) point out, 'Critics of biological determinism have frequently drawn attention to the ideological role played by apparently scientific conclusions about the human condition', and the nature of cognitive (and other) abilities described as 'natural' are not '"natural" but socially imposed obstructions to the building of a society in which the creative potential of all its citizens is employed for the benefit of all'.

In Chapter 6 I will return to consider how sets of assumptions persist, even when they are demonstrably dysfunctional to individuals and societies, because they form part of a pervasive 'worldview'. In the next chapter, though, I start by looking at assumptions about the nature of gene functions themselves.

References

Bateson, P. (1988). The active role of behaviour in evolution. In M.-W. Ho and S.W. Fox (eds) *Evolutionary Processes and Metaphors.* Chichester: Wiley.

Beckwith, J. (1987). Criticism and realism. *Behavioral and Brain Sciences*, 10, 72–73.

Beilin, H. (1987). Current trends in cognitive developmental research: towards a new synthesis. In B. Inhelder, D. de Caprona, and A. Cornu-Well (eds) *Piaget Today*. Hove: Erlbaum.

Benbow, C.P. and Lubinski, D. (1993). Psychological profiles of the mathematically talented: some sex differences and evidence supporting their biological basis. In *CIBA Foundation Symposium 178: The Origins and Development of High Ability*. Chichester: Wiley.

British Psychological Society (1991). The undergraduate curriculum in psychology: a first consultative report of a working party. Unpublished paper.

Burtt, E.A. (1932). *The Metaphysical Foundations of Modern Physical Science.* London: Routledge & Kegan Paul.

Cain, W. (1992). *Theories of Development: Concepts and Applications*. Englewood Cliffs, NJ: Prentice-Hall.

Carey, S. (1988). Are children fundamentally different kinds of thinkers and learners than adults? In K. Richardson and S. Sheldon (eds) *Cognitive Development to Adolescence*. Hove: Erlbaum.

Carey, S. and Gelman, R. (eds) (1991). *The Epigenesis of Mind: Essays on Biology and Knowledge.* Hillsdale, NJ: Erlbaum.

Casey, G. and Moran, A. (1989). The computational metaphor and cognitive psychology. *Irish Journal of Psychology*, 10, 143–161.

Chomsky, N. (1957a). Review of B.F. Skinner 'Verbal language'. *Language*, 35, 26–55.

Chomsky, N. (1957b). *Syntactic Structures.* The Hague: Mouton.

Chomsky, N. (1980) *Rules and Representations.* Oxford: Blackwell.

Cosmides, L. and Tooby, J. (1994). Origins of domain-specificity: evolution of functional organization. In L.A. Hirschfeld and S.A. Gelman (eds) *Mapping the Mind: Domain Specificity in Cognition and Culture.* Cambridge: Cambridge University Press.

Darwin, C. (1839). *Journal of Researches*. London: Colbourn.

Darwin, C. (1859). *On the Origin of Species by Means of Natural Selection.* London: Murray.

Dennett, D. (1978). *Brainstorms*. Montgomery, AL: Bradford Books.

Depew, D.J. and Weber, B.H. (1995). *Darwinism Evolving.* Cambridge, MA: MIT Press.

Detterman, D.K. (1993). Giftedness and intelligence. In *CIBA Foundation Symposium 178: The Origins and Development of High Ability*. Chichester: Wiley.

Dobzhansky, T. (1962) *Mankind Evolving: The Evolution of the Human Species.* New Haven, CT: Yale University Press.

Fodor, J. (1983). *The Modularity of Mind.* Cambridge, MA: MIT Press.

Fodor, J. (1985). Precis of *The Modularity of Mind. Behavioral and Brain Sciences*, 8, 1–42.

Gardner, H. (1984). *Frames of Mind: The Theory of Multiple Intelligences.* London: Heinemann.

Geary, D.L. (1996). Sexual selection and sex differences in mathematical abilities. *Behavioral and Brain Sciences*, 19, 167–228.

Gesell, A. and Ames, P. (1942). The ontogenetic organization of prone behaviour in human infancy. *The Journal of Genetic Psychology*, 56, 247–263.

Gelman, R. (1991). Principles and constraints. In S. Carey and R. Gelman (eds) *The Epigenesis of Mind: Essays on Biology and Cognition.* Hillsdale, NJ: Erlbaum.

Goodnow, J.J. (1984). On being judged 'intelligent'. In P.S. Fry (ed.) *Changing Conceptions of Intelligence and Intellectual Functioning*. Amsterdam: North-Holland.

Goodwin, B. (1985). Constructional biology. In G. Butterworth, J. Rutkowska, and M. Scaife (eds) *Evolution and Developmental Theory*. Brighton: Harvester Wheatsheaf.

Hall, A.R. (1983). *The Revolution in Science, 1500–1750*. London: Longman.

Hamilton, E. and Cains, H. (eds) (1961). *Collected Dialogues of Plato* (trans. W.K. Guthrie). Princeton, NJ: Princeton University Press.

Herrnstein, R.J. and Murray, C. (1994). *The Bell Curve*. New York: Free Press.

Hirschfeld, L.A. and Gelman, S.A. (1994). *Mapping the Mind: Domain Specificity in Cognition and Culture.* Cambridge: Cambridge University Press.

Hirst, W. and Manier, D. (1995). Opening vistas for cognitive psychology. In L.M. Martin, K. Nelson, and E. Tobach (eds) *Sociocultural Psychology: Theory and Practice of Doing and Knowing.* Cambridge: Cambridge University Press.

Ho, M.-W. and Fox, S.W. (eds) (1988). *Evolutionary Processes and Metaphors.* Chichester: Wiley.

Hobson, P. (1993) *Autism and the Development of Mind.* Hillsdale, NJ: Erlbaum.

Isaacs, N. (1966). Critical notice. *Journal of Child Psychology and Psychiatry*, 7, 155–158.

James, O. (1996). The Larkin Syndrome. *Observer*, 8 September.

Johnston, T.D. (1994). Genes, development, and the 'innate' structure of the mind. *Behavioral and Brain Sciences*, 17, 721–722.

Karmiloff-Smith, A. (1992). *Beyond Modularity: A Developmental Perspective on Cognitive Science.* Cambridge, MA: MIT Press.

Karmiloff-Smith, A. (1993). Self-organisation and cognitive change. In M.H. Johnson (ed.) *Brain Development and Cognition.* Oxford: Blackwell.

Keil, F. (1988). On the structure-dependent nature of stages of cognitive development. In K. Richardson and S. Sheldon (eds) *Cognitive Development to Adolescence.* Hove: Erlbaum.

Kellman, P.J. and Spelke, E.S. (1983). Perception of partly occluded objects in infancy. *Cognitive Psychology*, 15, 483–524.

Kimble, G.A. (1993). Evolution of the nature–nurture issues in the history of psychology. In R. Plomin and G.E. McClearn (eds) *Nature, Nurture and Psychology*. Washington, DC: American Psychological Association.

Kitcher, P. (1985) *Vaulting Ambitions: Sociobiology and the Quest for Human Nature*. Cambridge, MA: MIT Press.

Krebs, J.R. and Davies, N.B. (1981). *An Introduction to Behavioural Ecology*. Oxford: Blackwell.

Lewis, C. (1994). Episodes, events and narratives in the child's understanding of mind. In C. Lewis and P. Mitchell (eds) *Children's Early Understanding of Mind*. Hove: Erlbaum.

Lewontin, R.C. (1994). *The Doctrine of DNA: Biology as Ideology*. Harmondsworth: Penguin.

Light, P. and Perret-Clermont, A.-N. (1991). Social context effects in learning and teaching. In P. Light, S. Sheldon, and M. Woodhead (eds) *Learning to Think*. London: Routledge in assocation with The Open University.

Lloyd, B. and Duveen, G. (1991). The reconstruction of social knowledge in the transition from the sensorimotor to conceptual activity: the gender system. In M. Woodhead, R. Carr, and P. Light (eds) *Becoming a Person*. London: Routledge in association with The Open University.

Lumsden, C.J. and Wilson, E.O. (1981). *Genes, Mind and Culture*. Cambridge, MA: Harvard University Press.

McGraw, M.B. (1940). Neuromuscular development in the human infant as exemplified in the achievement of erect locomotion. *Journal of Pediatrics*, 17, 747–771.

Mayr, E. (1970). *Populations, Species and Evolution*. Cambridge, MA: Belknap Press.

Mehler, J. and Dupoux, P. (1994). *What Infants Know: The New Cognitive Science of Early Development*. Oxford: Blackwell.

Meltzoff, A.N. and Gopnik, A. (1993). The role of imitation in understanding persons and developing theories of mind. In S. Baron-Cohen, H. Tager-Flusberg, and D. Cohen (eds) *Understanding Other Minds: Perspectives from Autism*. Oxford: Oxford University Press.

Moore, C. (1996). Theories of mind in infancy. *British Journal of Psychology*, 14, 19–40.

Newson, J.E. and Newson, E. (1977). *Perspectives on School at Seven Years Old*. London: Allen & Unwin.

Oyama, S. (1985). *The Ontogeny of Information*. Cambridge: Cambridge University Press.

Piaget, J. (1929). *The Child's Conception of the World*. Totowa, NJ: Littlefield, Adams.

Plomin, R. (1992). Series editor's introduction. In T.D. Wachs, *The Nature of Nurture*. London: Sage.

Plomin, R. and Thompson, L.A. (1993). Genetics and high cognitive ability. In *CIBA Foundation Symposium 178: The Origins and Development of High Ability*. Chichester: Wiley.

Plunkett, K. and Sinha, C. (1992). Connectionism and developmental theory. *British Journal of Developmental Psychology*, 10, 209—254.

Rice, F.P. (1992). *Human Development: A Life Span Approach.* New York: Macmillan.

Richardson, K. (1990). *Understanding Psychology.* Buckingham: Open University Press.

Richardson, K. (1992). *Understanding Intelligence.* Buckingham: Open University Press.

Richardson, K. and Webster, D.S. (1996). Recognition of objects from point-light stimuli: evidence of covariation structures in conceptual representation. *British Journal of Psychology*, 87, 1–26.

Robinson, D.N. (1981). *An Intellectual History of Psychology.* New York: Macmillan.

Rose, S., Kamin, L.J., and Lewontin, R.C. (1984). *Not in Our Genes.* Harmondsworth: Penguin.

Scarr, S. (1992). Developmental theories for the 1990s: development and individual differences. *Child Development*, 63, 1–19.

Scarr, S. (1993). Biological and cultural diversity: the legacy of Darwin for development. *Child Development*, 64, 1333–1353.

Scarr, S. and Carter-Saltzman, L. (1982). Genetics and intelligence. In R.J. Sternberg (ed.) *Handbook of Human Intelligence.* Cambridge: Cambridge University Press.

Scholnick, E.K. (1994). Redescribing development. *Behavioral and Brain Sciences*, 17, 727–728.

Shepard, R.N. (1994). Perceptual–cognitive universals as reflections of the world. *Psychonomic Bulletin and Review*, 1, 2–28.

Silverman, I. and Eals, M. (1992). Sex differences in spatial abilities: evolutionary theory and data. In J.H. Barkov, L. Cosmides, and J. Tooby (eds) *The Adapted Mind.* Oxford: Oxford University Press.

Slater, A. (1989) Visual memory and perception in early infancy. In A. Slater and G. Bremner (eds) *Infant Development.* Hove: Erlbaum.

Spelke, E. (1994). Initial knowledge: six suggestions. *Cognition*, 50, 431–445.

Spelke, E.S., Breinlinger, K., Macomber, J., and Jacobson, K. (1992). Origins of knowledge. *Psychological Review*, 99, 605–632.

Sperber, D. (1994). The modularity of thought and the epidemiology of representations. In L.A. Hirschfeld and S.A. Gelman (eds) *Mapping the Mind: Domain Specificity in Cognition and Culture.* Cambridge: Cambridge University Press.

Sternberg, R.J. (1984). Towards a triarchic theory of intelligence. *Behavioral and Brain Sciences*, 7, 269–315.

Subbotsky, E. (1993). *Foundations of the Mind.* Cambridge, MA: Harvard University Press.

Thelen, E. and Smith, L.B. (1994). *A Dynamic Systems Approach to the Development of Cognition and Action.* Cambridge, MA: MIT Press.

Trevarthen, C. (1979). Instincts for human understanding and for cultural cooperation: their development in infancy. In M. von Cranach, K. Foppa, W. Lepenies, and D. Ploog (eds) *Human Ethology: Claims and Limits of a New Discipline.* Cambridge: Cambridge Univerity Press.

Trevarthen, C. (1983). Interpersonal abilities of infants as generators for transmission of language and culture. In A. Oliverio and M. Zappella (eds) *The Behaviour of Human Infants*. New York: Plenum Press.
Wellman, H.M. (1990). *The Child's Theory of Mind*. Cambridge, MA: MIT Press.
Wilson, E.O. (1975). *Sociobiology: The Modern Synthesis*. Cambridge, MA: Harvard University Press.
Wolpert, I. and Lewis, J. (1975). Towards a theory of development. *Federal Proceedings*, 34, 14–20.

2

GENE PARTNERS

Introduction

According to the assumptions and their sequelae described in Chapter 1, the structures and functions of cognitive ability originate in the genes. Even when interactions with the environment are referred to, the genes seem to retain their status as the head of a 'line management' in the assembly of human characters. These origins have been built up by selection forces acting on initially random variations. And the structures and functions within them are thus stable, fixed, adaptations to an equally stable set of environmental conditions (the accessibility and form of food; the activity of predators; particular kinds of physical problems, or social demands).

According to the almost universally accepted two-layer model, selection forces have acted to produce our knowledge structures and reasoning processes every bit as much as our physical appendages. Cognitive ability (and its variation) stands in relation to the genes exactly as do the structures and functions of the body. In the strongest accounts, the former are no more developmentable than the latter; and the genes produce individual differences in cognitive abilities in exactly the same way as in readily observable characters such as height or size of nose. The environment may have a supportive role, but it does not seriously affect form and variation. 'Feeding a well-nourished but short child more and more will not give him the stature of a basket ball player. Feeding a below average intellect more and more information will not make her brilliant' (Scarr 1992: 16).

In such accounts, then, our cognitive ability evolved to 'match' fixed, durable 'problems to be overcome' in the outside world. This is the simple adaptationist assumption picked up and embraced by psychologists. For example: 'Natural selection operates through the testing of alternative designs through repeated encounters with evolutionarily recurrent situations' (Cosmides and Tooby 1994: 86); 'Natural selection shapes domain-specific mechanisms so that their structure meshes with the evolutionarily-stable features of their particular problem-domains. Understanding the evolutionarily stable features of problem-domains – and what selection favoured as a solution under ancestral conditions – illuminates the design of cognitive specializations' (Cosmides and Tooby 1994: 96). As such domains Keil (1988: 97) mentions moral reasoning, spatial layout, social conventions, and so on, which 'have been of sufficient importance in evolution' to have led to 'specific predispositions' for cognition to occur in specific ways.

The point I shall be stressing in this and subsequent chapters is that this assumption is quite untenable for all but the simplest physical or physiological traits adapting to the simplest aspects of the environment. It is untenable because most aspects of the real world are constantly changing; rather than presenting 'recurrent situations', real environments tend to be in a state of dynamic flux. Not only is this characteristic of the physical principles of the world over time, but even the simplest organisms which utilise aspects of the environment alter them in the process. This is even more obvious among animals which behave (behaviour itself being an adaptable character in changeable conditions, as I shall explain further in Chapter 3). And it is particularly the case with human social worlds, in which social conventions, the grounds for moral reasoning, experienced 'spatial layout', and so on, may change drastically in the course of a lifetime.

The most prominent aspect of evolutionary history, indeed, is not one of a long succession of single 'lock and key' adaptations, but of new *systems of adaptability*. In changeable circumstances, many characters evolve which cannot be thought of as stable 'fixtures' of the organism in the way that most physical characters can:

> there are other characters of organisms and populations which are not explicable as adaptations to particular environments. . . . These and other traits may be regarded as adaptations to the pattern of the environment in space and time, to temporal variability, to environmental uncertainty. . . . When our emphasis shifts to variable environments entirely new problems arise.
>
> (Levins 1968: 10)

The irony is that it is just such adaptations-for-change that have lifted living things from bundles of fixed adaptations into the realm of cognitive abilities in the first place. Whereas Dawkins (1989: 13) argues that 'the differential survival of replicators [genes] is a special case of a deeper, more universal physical law governing the "survival of the most stable"', I shall argue that the world confronted by living things, especially humans, is governed by more complex laws, replete with *in*stability, and *for* which a purely genetic level of regulation is quite inadequate.

In these new, evolved levels of organisation, the two-layer conception of adaptation – genes on the 'inside', adapted characters on the 'outside' – is a quite inadequate one, either as an understanding of cognitive characters, *or* of diversity in them. In the course of evolution the organisation of adaptation has shifted from simple genetic regulations to genomic regulations, epigenetic regulations, cognitive regulations, or even, as in humans, *socio*-cognitive regulations – by which time the whole nature of 'adaptive problems' has been turned on its head because humans adapt the world to themselves, rather than vice versa.

The preoccupation with a single evolutionary story, in other words, has obscured another, far more thrilling, story. The latter not only calls into question many of the assumptions described in Chapter 1, but also totally alters our description of the role of genes, as well as the structure and function of our cognitive abilities. The task of this chapter and the next is to sketch out that story.

Genes and environmental change

Fundamental to the story is the nature and structure of environmental change. Of course it is a cliché of developmental psychology books that what we see in organisms and people is a result of 'genes *and* environments', even though the message that follows, explicitly or implicitly, is that the genes do the 'instructing' while the environment merely assists or attenuates. Even when psychologists speak of 'interaction', as they frequently do, the genes somehow manage to retain a commanding role. In consequence, descriptions of the environments relevant to cognitive abilities have remained in a very backward state. That environment tends to be conceived in very general, intuitive terms, such as the stable contingencies of natural selection, the general 'richness' of experience, the amount of physical or parental 'stimulation' in the home environment, and so on. Likewise, any change tends to be described in terms of isolated, independent factors, without awareness of any interactional structure in it.

In focusing on isolated, stable, aspects of environments, psychologists have followed adaptationists in general. The traditional adaptationist model has been more concerned with ideal types 'fitted' to stable environmental structures than the dynamics of a changeable world, except in a limited sense – in the sense that everyone now knows that there have been slow gradual changes in the world over long time periods, and that the succession of species over epochs is somehow related to them. But the model of the environment 'selecting' characters and thus the genes which produce them can only cope with change of a very superficial sort, namely slow change, in the same direction, over a very long time (Plotkin and Odling-Smee 1979). This is where 'pools' of genetic variability have always been seen as vital (Roff 1992). While there is reduction of genetic variability in *currently* adapted characters, important genetic variation persists in others, by continual mutations and reassortments. This 'allows' adaptability to future (changed) environments. By accidentally having genetic variants available for future selection contingencies, species have been able to 'track' environmental change, and thus persist in some revised form. Hence the traditional assertion that genetic diversity equals species vitality.

This phenomenon of currently redundant genetic variability, eventually becoming 'useful' in the face of environmental change, has often been called 'preadaptation' (Bock 1959). A good example is the presumed genetic variability present among the pectoral fins of lung-fishes in the Devonian period. When the pockets of water in which they had lived began to dry up some individuals had the (previously redundant) strength to convey their possessors across dry land, first as a means of reaching other pools, ultimately as a means of surviving on dry land. The new selection pressures, for what were then the genes of the few, 'weeded out' the old favourites, to build up a new genetic commonality: those survivors became the precursors of the first amphibians.

Even in some *selected* characters, genetic variability may persist, if, for example, the aspect of the environment adapted to is 'patchy' in time or space. In this case different subsets of genes underlying the adapted character are related to different forms or 'morphs' better suited to one 'patch' than to others. There are many examples of such 'genetic polymorphisms'. At the morphological (anatomical) level, they include variations in wing-coloration in moths which match polluted or non-polluted tree barks (Ford 1975). Evidence for polymorphic selection in variable circumstances is also shown in the shells of snails which have fallen prey to thrushes, tending to be those which are most distinct from their backgrounds: striped shells in woodland, plain shells in grassland (Clarke 1960). It is thought that genetic polymorphisms in humans, such as the different blood groups, or certain enzyme variations, may have been related to selection factors at some time in human history.

Modification and adaptedness by genetic selection is an essentially *transgenerational* regulation, but it remains, in numerous organisms, for numerous characteristics, an effective mechanism for life survival in changing environments. It has remained particularly suitable for those microscopic organisms which have very short generation times, like bacteria which reproduce, and thus create new variants, in the space of half an hour. Indeed, the rate at which they can produce new, antibiotic resistant strains, by natural selection of mutants, is an ever-increasing worry in our hospitals.

But this very point about short generation times hints at the big weakness of the process as a general survival strategy in more rapidly changing environments. The production of genetic variants is an essentially random process, and selection requires many generations for it to act, so the transgenerational strategy is only efficient with environmental change which is slow and 'directional' (i.e. continues to change in the same direction for a long time relative to the generation time of the species – Plotkin and Odling-Smee 1979; Plotkin 1988). (It is also wasteful: in the process huge numbers of inappropriate alleles have to be discarded.) The strategy has obviously worked for the long-term environmental changes across epochs, involving such variables as temperature, available oxygen, aridity, and so on. But many environmental changes will be both faster and more complex than this. Such changes may seriously threaten the existence of living things unless they, too, can be tracked in some way. This, in turn, requires new, *intra*-generational regulatory systems.

In sum, many adaptations *can* be identified as fixed structures or functions matching reliably recurring environmental conditions (c.f. Darwin's finches again). But it is a mistake to imagine that all adaptations are like this, and thus that the system of modification by selection of genetic variants is universally applicable. This is a fallacy, applicable only to special sets of conditions. Even Darwin (see the quotation in the Preface, p. xii) was sensitive to this problem. Real scientific problems start when we begin to use this model as a *general* model applied to all characters.

More generally, biological evolution can be seen as the struggle by life processes to overcome the challenges of constant change (Slobodkin and Rapoport 1974; Grobstein 1978). As organisms have evolved so they have brought about *accelerating* environmental change through their own activities. And as the environment has evolved it has become a more dynamic, ever-changing phenomenon. By definition, information-for-adaptation within a changing environment is only limited and temporary, and its enshrinement in permanent stores like genes would be self-defeating. Instead, new regulatory systems, which 'gain' information within generations, have had to evolve (Plotkin and Odling-Smee 1979; Plotkin 1994).

This requirement has resulted in all that we see as distinctive about advanced organisms, even in their genes. What I now go on to show is that, far from being independent 'beads on a string', serving as 'codes' for characters, and so on, genes have been incorporated in a regulatory system constituted by the genome as a whole; and further levels of organisation or regulation have evolved as organisms have confronted new degrees of environmental change. The result is a nested hierarchy of regulations in which the activities of lower levels have been incorporated in, and thus conditioned by, activities at the higher levels. In such regulatory systems, new, constructive processes of development have augmented direct gene expressions and gene diversity as the means of adaptation. In this chapter I consider one such system, and then go on in Chapter 3 to consider a number of others.

Genomic regulations

Until about twenty years ago, almost all the knowledge we had about genes came from comparisons of phenotypes. This consisted of statistical modelling of variation of phenotypes arising in breeding programmes in experimental plants and animals, and thus drawing inferences about underlying genes. A little more recently, mutations have been artificially produced in experimental organisms (e.g. by heat treatments of the embryos or exposure to ether vapours). Following such mutations, the consequences of 'deletions' of genes could be observed in the phenotype. Molecular genetic studies initiated only in the last two decades, though, have considerably enhanced our knowledge of associations between genes, their protein products and their ultimate consequences in the phenotypes.

The approach has consisted of a separation of proteins (the gene

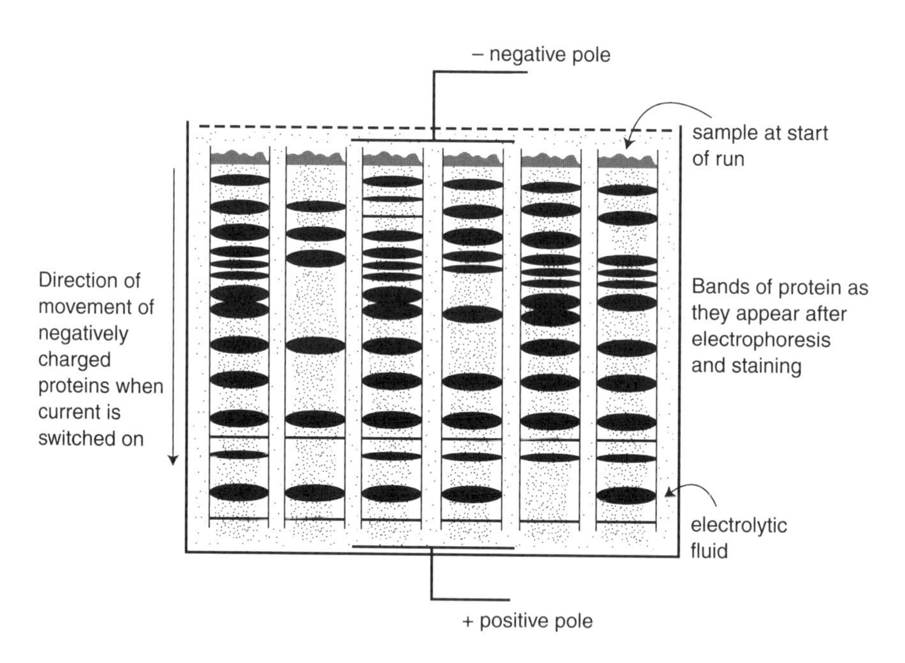

Figure 2.1 Diagram of proteins separated by electrophoresis. A homogenate of cells is placed at the top of the tube filled with gel. When the current is switched on different proteins migrate at different speeds and separate. Comparisons may indicate which genes are 'active', either in different individuals or the same individual at different times.

products themselves) through the technique known as electrophoresis. Different proteins tend to exhibit different electric charges. When extracts (homogenates) from tissues from an organism are placed on top of a tube or slab of a gelatin-like substance and an electric current passed through, the proteins are carried down the tube or slab at different rates. These *separated* proteins can then be revealed through staining, and their identities compared across extracts (Figure 2.1). This gives us a closer picture of gene products and thus gene identities.

Even with these techniques, almost all the knowledge we have about genes in the development of characters has come from the use of organisms like fruitflies, toads, and fishes, and of simple traits like body form, wing form, and so on. Still, they have produced a fascinating picture that has recently revolutionised our image of the gene. What follows can offer nothing more than the briefest glimpse of some substantive findings. I hope it is one, though, that conveys the general nature of this revolution.

Lessons from a fly

The fruitfly (or Drosophila) has been a most rewarding subject for geneticists for several reasons. It is commonly available, is easily kept in laboratories, shows, in its characters, clear responses to mutagens, has a short breeding cycle, and so on. But even flies are complex creatures, well adapted to many environmental fluctuations not found in the simple niches of bacteria, amoeba, and the like. One of these is overcoming the problem of developing a complex body in order to withstand the buffeting any embryo faces from external conditions. This, as we shall see, requires far more than a simple process of genes switching themselves on or off, or a direct assembly from genes to characters.

Fruitflies lay eggs which have been fertilised as in any other insect, and the development of the embryo begins almost immediately through rapid cell division. This process would, of course, simply produce an ever-growing ball of undifferentiated cells if it were not regulated in some way. So the first problem is how to start the process of differentiation into the beginnings of body parts – in fact, of the segmented larva known as the *instar* which has a head, three thoracic segments, eight abdominal segments and a tail region (Figure 2.2), and corresponding internal organs. And the first step of *this* problem is working out which is going to be the front and which is going to be the rear of the embryo (and, subsequently, the new fly).

The assumption of omnipotent genes as codes for characters might suggest that this involved the genes simply switching themselves on and issuing appropriate instructions for the building of different body parts. Since all cells contain exactly the same genes, however (i.e. they are ‘multipotent’), this would merely produce an ever-bigger ball of

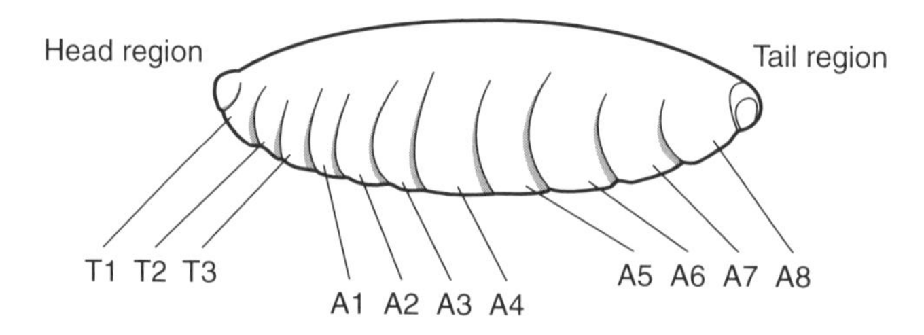

Figure 2.2 General impression of segmented Drosophila larva about twelve to fourteen hours after fertilisation (T = thoracic segments; A = abdominal segments). Both head and tail regions house internal structures that are far more complex than their outer appearance suggests.

identical cells. So some further discriminatory process is obviously required. But how is this discrimination achieved?

Fortunately, contrary to popular impressions, offspring inherit rather more than their genes from their mothers. They also receive the whole complex machinery of a cell – a host of active biochemicals, membranes, and organelles. Moreover, they inherit the products of numerous *maternal* genes, left in the cell after the egg has been laid. It appears to be *these* products – not the embryonic genes – which orchestrate the first developmental steps (Finkelstein and Perrimon 1991).

However, this process, too, is more than a simple 'switching on' or 'switching off' of structural genes. As their name implies, the structural genes produce the structural and enzymatic proteins manifesting the body form; but the 'where' and the 'when' of these activities is not determined by them. For example, the product (in fact messenger RNA, or mRNA, see Chapter 1, p. 15) of the maternal gene known as *bicoid* appears to be critical in the formation of the embryonic head. It is present in the egg, but, instead of being spread around evenly, it has its distribution constrained by the products of two *other* maternal genes. This uneven distribution, or gradient, from front to back, of the bicoid product seems to be important in deciding which of the structural genes are recruited, and where and when, thus influencing eventual head and tail differentiation. Mutations deleting the bicoid gene and its product in the mother (and therefore its presence in the egg) results in the deletion of the entire head and thoracic structures in the embryo. They are replaced, though, not by a zone of undifferentiated cells, but by duplicated *tail* structures. This suggests the mediation of still another 'raft' of regulatory genes influencing what the structural genes actually produce, and where and when (Finkelstein and Perrimon 1991).

A number of gene systems dependent on the concentration gradients of maternal gene products have been identified (Finkelstein and Perriman 1991; Edgar 1992). Regulations similar to those just described have been observed to operate for other aspects of body form, such as the differentiation of the ventral (underside) and dorsal (back) of the embryo (Morisato and Anderson 1995). In each of these a complex hierarchical system of regulation is observed. Moreover, it appears that the same genes can be used *differently* depending on where and when they operate in the front–back, or top–bottom axis.

For example, the gene-product known as 'Dorsal' protein becomes distributed in the way just mentioned for the bicoid product. But this gradient then determines which of a number of *other* key regulatory genes involved in tissue differentiation actually become expressed and when. Its role then varies, depending on the presence of other factors around those genes. 'By using Dorsal as both a transcriptional activator and repressor the embryo uses the information contained in the Dorsal gradient to regulate gene expression along the entire axis. Dorsal is inherently an activator, but can mediate repression by interacting with additional factors bound to neighbouring sites' (Morisato and Anderson 1995: 385). Of course, the point is that the expression of the ultimate, structural genes in a character only emerges as the result of complex interactions among a whole team of genes.

Another interesting case of this regulatory interplay involves the so-called *ultrabithorax* (*UBX*) gene. This gene appears, from mutation studies, to be involved in the development of the different kinds of distinct body segments of which flies and their embryos are made (i.e. thoracic segments, abdominal segments; see Figure 2.2). It used to be thought that each kind of segment would be targeted by specific structural genes simply switching on and producing the proteins they code for. But the system is both more regulated and more flexible than that. It turns out that the specification is ordered by regulatory genes acting such that 'a single [structural] gene can specify more than one segment type' (Castelli-Gair and Akam 1995: 2973).

This is illustrated in the way that the spatial distribution of the *UBX* protein varies across segments in different concentrations at different times, with different consequences for the expression of a structural gene, called *Antennapedia* (*Ant*). As a result, the development of certain structures in some segments is repressed where, at that time, the *UBX* protein concentration is initially strong, but not in others in which its concentration is weak. By the time *UBX* distribution has become more even, the primordia of structures already forming are no longer repressed by *UBX* protein. Thus, what the structural genes will do 'depends on the context in which they are expressed, and need not always have the same consequences in terms of "segment identity"'(Castelli-Gair and Akam 1995: 2981).

Here, again, it is obvious that complex interactions ensue in which products of some genes are switching on other genes, arranging the distribution of their products, and inhibiting still others in a time-varying sequence which results in the differentiation and proliferation of quite different cell types and tissues. No single 'command' genes can be said to *determine* this production and differentiation. Rather, it is the result of the interactive processes in a self-organising sequence over time, the outcomes of one interaction determining the nature of the next, and so on.

The importance of such a regulatory system, maintaining development in the face of environmental buffeting, is seen starkly in another sense. It may be thought that the regulatory genes may be small in number compared to the structural genes, which actually do the business of producing structural proteins. In fact the opposite is emphatically the case: it is now estimated that, even in the fruitfly, *up to 90 per cent of genes are involved in regulation of other genes*, rather than in the direct building of body parts (Lawrence 1992).

This elaborate, but self-regulating, orchestration is, of course, a far cry from the expression of independent, separate loci, as in the 'standard model' of gene action. Mechanisms have clearly evolved in which the expression of structural alleles is incorporated into regulations at a higher hierarchical level (Rollo 1995). Even at the level of a fly, what offspring inherit from parents is not simply genes with 'instructions' to build another being directly, but a *developmental system*. Another important aspect of this system is the existence of 'regulatory regions' on genes. These are not minor adjuncts to what might otherwise be thought of as the real business, or coding, part of the gene. '[I]n most important genes involved in development, their regulatory regions are enormous compared with their protein-coding regions. In [a typical] complex, 95% of DNA is regulatory and only 5% is transcribed' (Rollo 1995: 92).

These regulations have evolved, of course, in order to 'buffer' development against the violent environmental disturbances experienced by an embryo, and thus maintain development on a reliable track. In the process, of course, the activities of genes in cells become aspects of the environment of other genes and cells. This is important for further differentiation. The fact that all cells contain a full complement of genes (i.e. are 'multipotent'), yet differentiate into numerous different cell types, itself testifies to the role of cell–cell interactions and other environmental cues. (Cells cultured in a plate in the laboratory do not 'automatically' differentiate in this way.) These are seen strikingly in observations of early development of brain structures.

As observed in amphibians and vertebrates, the brain originates as an infold of the outer layer of cells to form the 'neural tube' (Figure 2.3).

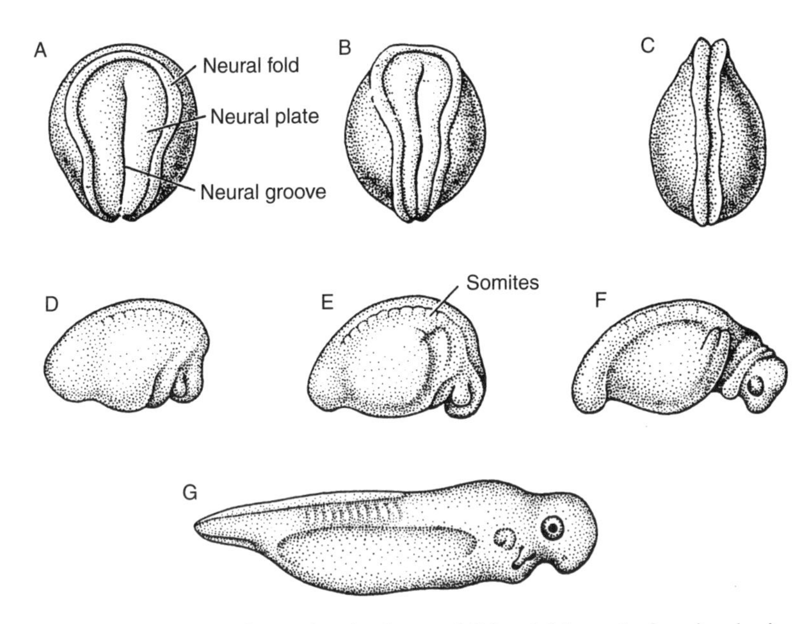

Figure 2.3 Neural tube formation in the amphibian (ultimately forming brain and spinal cord) involves widespread migration and differentiation of cells and 'social' regulation of genes. (The neural folds forming in A–C eventually fuse in the midline (D–F) to form rudimentary brain and spinal cord (G).) From Purves and Lichtman 1985; reproduced with permission.

Cells within the neural tube subsequently migrate away from the tube, some of them travelling considerable distances to form a wide array of types of cells. These include sensory neurons for the sense organs, the various 'internal' neurons of the nervous system, the ganglion and other cells in and around the spinal cord, and the glia cells which interact with, and pack around, them. How can such widely different cells and tissues arise from a single type?

Again, this is not due to instructions arising directly from genes 'for' particular kinds of cells or their attributes. Rather, 'the cells differentiate according to instructive cues from their surroundings, acquired either along their migratory pathway or at their final sites of localization' (Fraser and Bronner-Fraser 1991: 913). This dependence of what genes do, *inside* cells, on complex regulations impinging from *outside,* has even been described as 'social control' (Philipson and Sorrentino 1992). Intensive studies recently have indicated further the mechanisms by which the products of some genes in some cells enter other cells and 'direct' the expression of genes therein, to influence the future course of the latter's development (see contributions in Russo *et al.* 1992). As already mentioned, this level of regulation is so crucial that 'most normal cells are

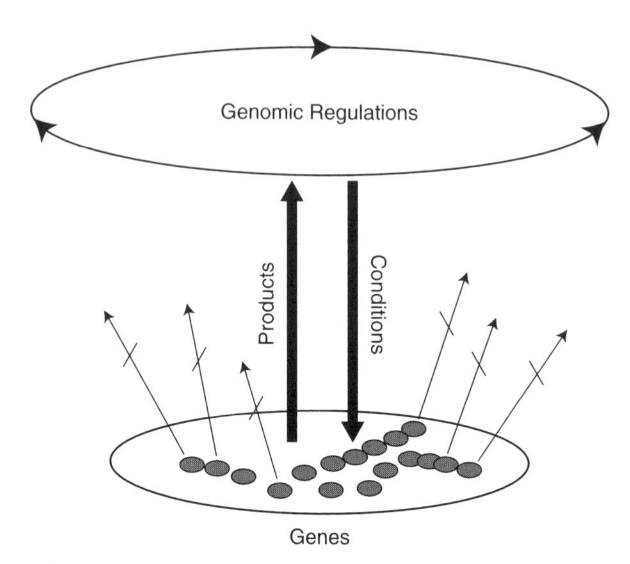

Figure 2.4 Rather than independent 'expressions' (arrows cancelled), production from genes has become organised within a higher, genomic level of regulation.

unable to divide unless they are anchored to an extracellular matrix, or . . . with other cells from which they receive signals' (Philipson and Sorrentino 1992: 539).

It is becoming increasingly clear, then, that genes, at least those associated with evolutionarily important characters, are not simple codes, mechanically producing their separate contributions to characters. Rather, these have become embedded in higher levels of genomic regulations (Figure 2.4). The forms that emerge arise in interactions between these levels, each setting conditions for the products of the other. Appropriate developmental end-states can thus be attained in the face of environmental perturbations, which may be severe. There is causal integration across numerous levels in which 'original' causes lose their identity. Ultimately, such nestedness of regulations, in which activities at one level are incorporated into regulations at a higher level, goes much, much deeper.

A new view of the gene

This elaborate, but self-regulating, orchestration is, of course, remote from that traditional view of expression of independent, separate loci, as in the 'standard model' of gene action. For critical characters, mechanisms have clearly evolved in which the expression of structural alleles is incorporated into regulations at a higher hierarchical level. This already implies a new view of the genes themselves, and several new aspects have emerged in recent years. One of these has been a more emphatic distinction between regulatory genes and structural genes. Another involves the existence of huge 'regulatory regions' on genes themselves (as mentioned above, up to 95 per cent of DNA may be regulatory and only 5 per cent structural).

As a result we have to move from the popular conception of genes as independent codes on strings to one of cooperative players in a complex hierarchical web of regulation. The simplest model of gene regulation, and one sometimes referred to in psychobiology texts as the 'standard', is that of an 'inducible' gene (one which is switched on or off by the level of a specific substance in its environment); for example, the *lac operon* which was described in the gut-dwelling bacterium *E.coli* by Jacob and Monod (1961). In this model, the gene producing the enzyme which metabolises milk sugar (lactose) is turned 'on' by the presence of lactose in the cell surround, and 'off' when it is absent. The mechanism is such that the lactose binds to a 'repressor' molecule, itself the product of a regulatory gene, in the bacterium. This binding stops the repressor protein from binding to an 'operator sequence' where it normally inhibits transcription of the gene which, in turn, codes for the enzyme in question. With such inhibition relieved, the enzyme is now produced and the milk can be utilised. When the milk dries up, the opposite sequence prevails, the repressor now binding to the operator sequence and preventing transcription and enzyme production (Figure 2.5).

Although a good model of regulation, the regulatory systems, and consequently the genes, of advanced (eukaryotic) species are much more complex. There are many more operator-type sequences (e.g. 'transcription factors', 'enhancer', 'silencer', and 'promoter' regions); they have larger regulatory regions for interacting with other distantly located genes; and the DNA is arranged around bodies of protein such that it (the

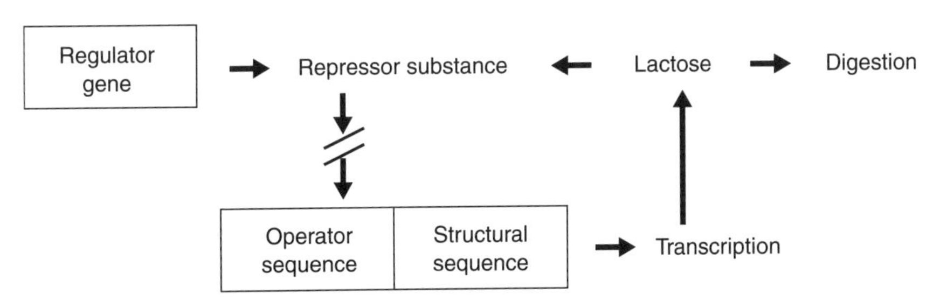

Figure 2.5 In the absence of lactose in the environment the product of the repressor gene binds to the operator sequence and inhibits transcription of the structural gene. Lactose lifts this inhibition by binding to the repressor substance, so yielding the enzyme that results in its digestion (after Jacob and Monod 1961).

DNA) may be available for transcription at some times and not at others, with transition between these states *also* under regulatory control (Grunstein 1992; Rollo, 1995).

A *very* simple description of this more complex process is as follows (Figure 2.6). First a specific set of regulatory proteins collects on the first section of a 'promoter' region to form a complex of the proteins and the DNA (Figure 2.5). This complex includes RNA polymerase, an enzyme which, as its name suggests, catalyses the formation of chains of RNA molecules. This enzyme then passes to another section of the promoter region called the initiation complex (I). From this position it is brought into contact with the coding region (the section of chain actually 'loops over' to achieve this). It then travels, like a train on railway lines, up and down the coding region of the DNA string, 'reading off' the coding sequences and matching them with corresponding units in another form of RNA – messenger RNA. This is what serves as the 'template' from which structural proteins and enzymes are constructed elsewhere in the cell. This whole process, however, produces only a basic or 'basal' level of the mRNA. Binding of still other regulatory proteins to another region – the *activator* or *enhancer* sequence – results in further activation of the promoter, and a great boost in mRNA and protein production.

As already implied, the regulatory system is actually much more complex even than this. It has become clear in the last decade or so that the promoter and coding regions tend to become arranged as a sequence of tiny coils around proteins called histones to form a 'nucleosome'. The histones were once thought to be merely packing material, but, in the nucleosomes, are now known to have crucial regulatory functions as well. Under (direct or indirect) induction from the activator proteins the histones 'unwind' their coils to expose the coding and promoter sequences they otherwise conceal. Thus the first step in turning a gene 'on' is the lifting of the histone repression. But it is suggested (Grunstein, 1992) that

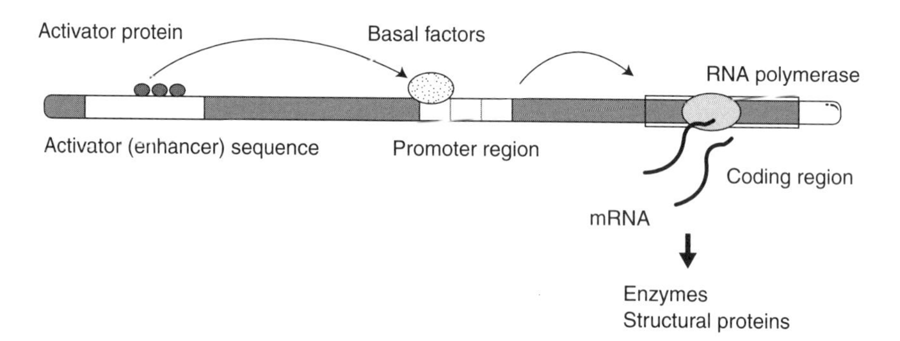

Figure 2.6 A much simplified model of the gene regulatory system in complex organisms (redrawn from several sources).

this occurs by *other* activator proteins binding to a regulatory segment of the histone complex, resulting in a change in its structure and some uncoiling of the nucleosome. After passage of the RNA polymerase, as described above, the histone complex may then be reconstituted.

We only have to realise, of course, that all these 'other proteins', involved in both the activator and promoter sequences, as well as the neucleosome complex, arrive on the scene as the products of other genes, to appreciate the extent of the web of regulations in which the gene is incorporated. The nature of this regulation is not always one of merely switching on or off. As already hinted above, it also appears that some regulatory genes may operate so as to allow *different* transcripts to be obtained from a single gene (Rollo 1995; Slack 1991; Lawrence 1992).

More startlingly, it has recently become clear that the very DNA sequence, or 'code', of at least some genes can be modified *by* the developmental processes. It is now known that genes are arranged on the chromosome in 'coding' stretches known as 'exons' interspersed by non-coding regions known as 'introns'. It appears that mechanisms exist for the rearrangement of the overall gene structure by 'exon shuffling', resulting in novel 'codes', producing novel proteins with novel properties, i.e. an ability for genomes to 'genetically engineer' themselves 'in a much more efficient manner than envisioned by the traditional view of random mutations' (Rollo 1995: 134; see also contributions in Ho and Fox 1988; Landman 1991; Wills 1991).

Thus, Ho and Fox (1988: x) speak of 'the discovery of the "fluid genome"' in which physiological processes affected by environmental perturbations result in changes *in genes themselves* which are then passed on to the next generation: Lamarckism, or the inheritance of acquired characteristics, long ridiculed by geneticists, has made a respectable empirical come-back (Landman 1991). As Rollo (1995: 80) explains, 'Probably the greatest new development in our view of the genome is the diversity and action of transposable elements whose activities dice and splice the genome. . . . Mobile elements have the capacity to rearrange cytogenic organization . . . as well as to modify the structure of genes themselves'.

Such processes of genetic self-rearrangement can lead to both diversity and 'extraordinary homogeneity' in the form of gene complexes which can then operate in concert (Depew and Weber 1995). Originally scattered genes associated with a complex trait can be transposed into nearby clusters which have come to be called 'supergenes', and thus transmitted from parent to offspring intact. These can evolve rapidly together and, by the sorts of mechanisms just mentioned, become standard in all members of a species 'without passing through Mendel's Laws' (Depew and Weber 1995: 352). These genomic regulations are now themselves seen as an 'adaptation', accelerating evolution to match rapid environmental change (Wills 1991; cf. Depew and Weber 1995).

This picture, too, is a far cry from the fatalistic model of genes, and their autonomous actions, presented in psychology texts. When we talk about genes being switched 'on' or 'off', this, in itself, involves many regulatory steps and the cooperation of many other genes and their products at several hierarchical levels. Rollo (1995) points out how knowledge of the complexity of organisation of the genome transforms the traditional view of genes mapping to the phenotype in a one-way one-gene-one-enzyme framework into one of complex integration and feedback across multiple levels.

The techniques of molecular biology, therefore, are progressively dispelling the notion of beads on a string or independent beans in a bag. 'Such dynamics suggest that the genic beanbag is an impossibly simplistic model. . . . As we scrutinize the molecular landscape with increasing sophistication, rather than discovering a lower level of fundamental control, regulation appears instead to recede rapidly into higher levels of organization' (Rollo 1995: 80). 'Ironically, what is being uncovered at the molecular level is not some fundamental level of invariant, ultimate control, but complex interactive systems with cascading levels of organization and numerous feedback systems' (Rollo 1995: 6).

An important implication of this modified view of the gene is that much phenotypic change, variation, or even identity can be achieved without changes in the genes as such, but by regulation of the timing and location of recruitment of existing genes. Genes originally or primarily subserving one purpose 'can be commandeered for other purposes with no loss of their original role' (Rollo 1995: 105). In the context of interactions between regulatory levels, genes become the *tools* of a developing, functioning system, rather than vice versa.

The fact that more or less independent genes have evolved in such regulatory directions is also implied in comparisons of amounts of DNA across species. Widely divergent species often show remarkably little diversity in their DNA. For example, although modern apes and humans split from a common ancestor millions of years ago they continue to share at least 99 per cent of their DNA. Or, in other words, what makes humans vastly different from apes, especially cognitively, involves only slightly different genes, and it is thought that these are largely regulatory rather than structural genes. Numerous genes involved in basic physiological processes indeed have homologies across phyla from, say, flies to mice to apes and humans. Their main evolutionary differences lie in the regulatory mechanisms through which DNA is recruited in development. Mice have six thousand times as much DNA as bacteria, which is certainly consistent with differences in complexity. But humans have *no more* DNA than mice. The clear implication is that the gradual emergence of behavioural complexity within the mammals was not achieved by accumulating 'better' single genes, but by increased regulation of those already available (Bateson 1988: 201).

The whole striking function, and evolutionary 'achievement', of these new regulatory levels become lost by a gene-centred, reductionist view. It is such developmental *systems* that have become the sources of complexity in evolution, not independent genes *à la* the central assumptions of adaptationism and selectionism. What come to be paramount in the survival of living things are no longer single genes but whole developmental systems. Biologists and psychologists now stress the idea of hierarchies and systems as the unit of adaptation, not isolated characters (Plotkin 1988, 1994; Depew and Weber 1995). By the same token, what we inherit from parents are not merely genes, not preformed abilities, but a system that is 'developmentable'. When we turn later to consider further levels of regulation, we shall see that this idea has important implications for our understanding of cognitive ability in humans.

There are, of course, numerous evolutionary implications of the picture now rapidly emerging in developmental molecular genetics, and traditional Darwinism is under something of a siege (see, e.g., Goodwin 1985; Ho and Fox 1988; Depew and Weber 1995). What seems particularly ironic at present is that just as psychologists are increasingly seeking to anchor their ideas in traditional Darwinism, biologists are moving into a 'post-Darwinian' period (Bocchi 1991). The main implication for this book, though, is that it is regulatory mechanisms, at increasingly higher levels, that have evolved, rather than the chance emergence of 'new' autonomous genes determining 'all that we are', and individual differences in it. I hope it will become clear, in the next chapter, that these transform organisms, and especially 'knowing' and 'thinking' humans, from passive entities under the rule of random genetic and environmental variation to far more active beings in the world.

Implications for the rhetoric of genetic determinism

The metaphors of genetic control, determination, programmes, etc. are widely used in both biology and psychology, but they are now clearly one sided. As Nijhout (1990: 441) points out, 'Although these metaphors have an admirable sharpness and punch, they lead, when taken literally, to highly distorted pictures of development as a process'. Yet it is undoubtedly true that such distorted pictures have dictated research priorities, funding grants, and even the terms in which academic papers have to be couched before they are deemed suitable for publication.

Nijhout (1990) has analysed the bases of the popularity of these metaphors. One is undoubtedly the consequences of mutations in which changes in genes can lead to often quite dramatic changes in gene products, and devastating sequelae in organic and psychological functions. Indeed, much of the powerful mystique surrounding genes seems to derive from such cases, normally found in a very small percentage of human populations. If the deconstruction of a character occurs when a gene is absent, surely the same gene can be said to 'cause' the construction of the same character when it is present?

But this is an error of logic. Rose (1981) has warned how it may foster the search for simplistic single causes, but, by crossing levels of regulation, results in grossly erroneous conclusions. The point is that lower levels of regulation may be said to furnish the organic and structural necessities for regulation at successively higher levels, without determining what those regulations actually consist of. Alterations or deletions in certain pieces of the furniture may, of course, disrupt what goes on at that higher level, but this is different from saying that they 'cause' what normally *does* go on. By the same token, we might as well argue that because a puncture brings a bus to a halt it is 'air' that causes the bus to move.

We now know that the rhetoric of genetic determinism is quite wrong. Genes by themselves do not 'make' proteins, 'self-replicate', 'control', 'constrain', 'programme' or issue 'instructions' for anything; neither do they 'express' themselves. Each of these things only happens through the resolutions of complex regulatory processes in the cell, themselves responding to extracellular and other interactions at and across numerous levels. Genes are best thought of as *resources* utilised by a dynamic system in a regulated manner. What we inherit from parents is not just a set of genes, and not a genetic programme, but a whole *developmental system* which utilises genes as resources.

So it is highly misleading to say that genes 'cause' development, or are codes for direct mapping into characters. A more accurate view is that the genes are recruited in the making of products which are necessary for development. In the context of interactions between regulatory levels, genes become the tools of a developing, functioning system, rather than vice versa (Rollo 1995). Although it has been commonplace to speak of the genotype as creating the phenotype, the kinds of interactions described above suggest that it is at least as legitimate to describe the developing phenotype *as creating its own genotype* (Oyama 1985). Rather than 'gene gods' it is better to think in terms of 'gene partners'.

The genomic regulations described so far are those involved in the attainment of crucial aspects of body form against hostile perturbations in the environment in which any embryo finds itself. These perturbations will involve changes in temperature, humidity, bodily orientation, and nutrients, as well as physical bumps and shocks. An arrangement of independent 'codes' would be extremely vulnerable in such circumstances, with many 'misfires', wasted products and wasted embryos. So that the sort of genomic regulations already discussed extend, as it were, the life and range of contribution of the genes, with new 'information for development' being found in the structural relations of product gradients, cell surrounds, and so on. In Chapter 3 I hope to show how such genomic regulations have themselves been opened up by a further, 'epigenetic' level of regulations, in which they have become embedded. And these epigenetic regulations have, in turn, been 'opened up' by cognitive and then socio-cognitive regulations in which they have, likewise, become embedded. All of these have to do with adaptations in increasingly changeable and difficult-to-predict environments, in which the information for survival has to be found at increasingly deeper structural relations among environmental variables.

A note on canalisation

All of the regulatory activities of the genomic system in the face of an unstable environment, discussed above, are those of the production of stable body parts. So 'tightly' is the process maintained that it has sometimes been called 'canalisation'. Canalisation refers to workings of the interactive system between genes, other products of development, and the environment, in maintaining development along a tightly specified pathway. It is worth discussing this topic here because it is one which has become very popular in developmental psychology in the last decade or so.

The present use of the term 'canalisation' derives from the developmental geneticist C.H. Waddington (1940) who claimed that, for many characters at least, development of the normal phenotype is so strongly buffered against even extreme environmental perturbations as to be maintained on its chreod, or 'fated' pathway. He reached this conclusion from the observation of certain courses of development into well-defined types, rather than a range of grades, in spite of varying genes in the embryo and/or environments of development. He could see that this was not due to genes alone, but to 'a developmental process whose course is steered by the combined action of the whole genotype and the impinging environment' (Waddington 1975: 6).

Rendel (1967: 42) stressed how the predictable course of development is maintained even when different genes are present (or have been introduced by mutation) among members of the population displaying the character: 'The uniformity of a population with respect to a canalised character is due to something more than the absence of genetic differences between its members'. Or, conversely, even when certain 'factors', genetic or otherwise, are missing, developmental processes persist because the integrity of the developing system is maintained through the multiplicity of those factors. Only when environmental experience extends beyond a certain range does the system 'break down' and underlying genetic variation become manifest.

Waddington described canalised development in terms of a ball rolling down a valley with its tendency to be diverted along side-valleys being suppressed by gene responses. Thus he concluded that the process is best thought of as a developmental manifold involving interactions between comparatively large numbers of genes (Waddington 1957: 131). There were numerous illustrations of apparently canalised development long before the recent molecular-genetic studies mentioned above. For example, mutation studies have shown that numbers of scutellar bristles in the fruitfly appear to be 'so tightly canalised that even a considerable amount of genetic substitution will not result in a visible change in the phenotype' (Mayr 1970: 174).

Gottlieb (e.g. 1991) has criticised Waddington's (and others') notion of canalisation on the grounds that it harks back to an implicit genetic control. It is not entirely inconsistent with a broader 'systems' view of development, though (a view for which Gottlieb has long been an advocate). I turn to that view in the next section. Furthermore, in a hierarchically embedded system of regulations it is quite conceivable for there to be tightly canalised development at one level furnishing the requirements for a far more open development at a higher level in which it is embedded. These are the subject of the next chapter. In it I shall be at pains to point out that canalisation is certainly *not* a useful description or metaphor for the development of cognitive abilities.

Dynamic systems theory

Attempts to oppose reductionism and 'mechanicism' in science have commonly consisted of appeals to organised *structures* and *systems*, involving ideas like transformations and self-regulation, rather than conglomerations of independent controlling factors. Piaget (1971) notes how the biologist L. von Bertalanffy long ago attempted to develop a 'general theory of systems' based on a structured 'organicism' (Bertalanffy, 1968). Such efforts have continued in recent times. The view of genomic *systems* just discussed, as well as other regulatory systems to be discussed in the next chapter, is, however, part of a more general shift in scientific theorising in recent years, which has become known as 'dynamic systems theory'. It is a view which eschews reductionist, monocausal schemes for modelling complex systems and, instead, attempts to capture 'the mutual dependence of causes . . . an entire ensemble of influences' (Oyama 1985: 15).

In many ways it is a reaction to the whole 'metaparadigm' governing science in the last two centuries, whereby causes and effects have been sought, and described and applied, in terms of a single, or a few, 'push–

pull' linear relationships. Although that approach has obviously been enormously beneficial, its limitations are being increasingly realised: nature is mostly far more complex than can be described in functional equations involving only a very few linear factors. Natural phenomena are more complex in the sense that they tend to consist of many variables, mutually dependent on each other, often in non-linear relationships, conditioned by others at a 'higher' level, the whole system often changing or evolving over time. If this is true of physical and physiological systems, it is likely to apply even more so to the most complex phenomena that exist, namely human cognitive abilities.

Dynamic systems theory demands a radical change in the way we think, scientifically, about phenomena. As Fraser (1988: 1) put it, 'Ten years ago if an experimenter obtained a complicated signal . . . not composed of a few simple peaks the common interpretation would be that there was noise in the signal or that the observed phenomenon had too many degrees of freedom to be reasonably modelled'. In the last few years new mathematical approaches have shown how the behaviours of complex inanimate and animate systems might be understood as self-organised, interactive systems. As Fraser (1988: 1) goes on, 'This picture is an appealing explanation of complicated behaviour because it is completely self-contained. There is no need to search for an external noise source (. . . reminiscent of the Aristotelian search for a prime mover)'. Complexity is seen to arise out of the interactions within the system itself.

Illustrations abound in the literature of how such interactions can be self-organised and present patterned coherent form and patterned change. The torrid flow of a mountain stream presents a pattern according to the myriad interactions between the force of gravity, the shape of rocks and banks, and so on, but the pattern cannot be said to have a single, ultimate determiner (Thelen and Smith 1994). Likewise, the pattern may *change* by erosion of the bank or displacement of a rock, but this change cannot be said to have a single, ultimate determiner either – although an understanding of the overall *dynamics* can help us to intervene (to prevent erosion, say). Somewhat similarly, the intricate pattern of bipedal walking can be shown to 'emerge' out of the interactions of bones, joints, and muscles and their inherent constraints and freedoms, rather than being dictated by a precise neural programme (though simple changes in neural inputs may alter the rate and/or pattern of interaction, as in the transition from walking to running) (Thelen and Smith 1994).

Dynamic systems theory can account for stability as well as change. Some of the patterns which emerge from the interactions may, given certain recurring conditions, be very predictable and ultimately stable because the system has reached an intrinsically 'preferred' state. Goodwin (e.g. 1988) and others have long argued that the development of body form in animals – morphogenesis – reflects such self-organising interactions arriving at preferred states rather than the direct 'intentions' of genes. That preference does not reflect a single creative agent, nor a determinate case. 'Although some behavioural preferences are so stable that they take on the qualities of a developmental stage, the stability is a function of the organism-in-context, not a set of prior instructions' (Thelen 1995: 100). Thus the appearance of an ability as 'innate' merely reflects such an equilibrium state (one which is theoretically moveable to another state). Development is then seen as the transition from one such preferred state to another, as one or other factors in the system (internal or external) changes and creates a disequilibrium.

It is too early to tell whether this approach will yield better theory about human abilities, or whether it will fizzle out through lack of theoretical detail. An obvious danger is that dynamic systems theory could simply present us with yet another set of shadowy metaphors for development and ability. While presenting appealing new principles at a general, abstract level, analyses of real behaviour in dynamic systems terms have so far been largely confined to fairly simple motor actions. There have been very few attempts to apply the perspective to actual cognitive abilities as yet, although in recent years I have proposed a theory of cognition based on 'covariation hyperstructures' that has many of the qualities of a dynamic systems theory (e.g. Richardson 1992; Richardson and Webster 1996a, 1996b).

But more of that in Chapter 3. In this chapter I hope to have at least begun to question some of the staple assumptions of nature–nurture debates in psychology. It is misleading to talk of any characters as 'innate', in the sense of preformed by some sort of singular active agent. It is also misleading to describe genes as ciphers or codes for characters. The adaptationist assumption is naive in changeable environments in which regulatory *strategies*, rather than fixed structures and functions, have to evolve (of which more in the next chapter). The genes associated with evolutionarily important characters will tend to become associated in cooperative ensembles and regulatory hierarchies, rather than persist as more vulnerable, independent 'codes'. This results in reduced, if not the complete disappearance of, genetic variability for such characters, in at least two senses. First, natural selection of key alleles itself tends to eliminate genetic variation. Second, in the context of genomic (and other) regulations, possibly even wide genetic variation does not matter, because variable genetic *means* can be utilised in the attainment of a common functional *end*. Recent research suggests that the mechanisms for this may even include modifications of genes themselves 'on line', as it were, in the developmental process.

Finally, the kind of regulatory framework just described should also dispel the notion of a 'bell-shaped' curve – an 'accidental' product of variable, independent genes – as a suitable model for evolutionarily important characters, especially cognitive ones.

References

Bateson, P. (1988). The active role of behaviour in evolution. In M.-W. Ho and S.W. Fox (eds) *Evolutionary Processes and Metaphors.* Chichester: Wiley.

Bertalanffy, L. von (1968). *General Systems Theory*. New York: George Braziller.

Bocchi, G. (1991). Biological evolution: the changing image. In E. Laszlo (ed.) *The New Evolutionary Paradigm.* London: Gordon & Breach.

Bock, W.J. (1959). Preadaptation and multiple evolutionary pathways. *Evolution*, 13, 194–211.

Castelli-Gair, J. and Akam, M. (1995). How the Hox gene *Ultrabithorax* specifies two different segments: the significance of spatial and temporal regulation within metameres. *Development*, 121, 2973–2982.

Clarke, B. (1960). Divergent effects of natural selection on two closely-related polymorphic snails. *Heredity*, 14, 423–443.

Cosmides, L. and Tooby, J. (1994). Origins of domain-specificity: evolution of functional organization. In L.A. Hirschfeld and S.A. Gelman (eds) *Mapping the Mind: Domain Specificity in Cognition and Culture.* Cambridge: Cambridge University Press.

Dawkins, R. (1989). *The Selfish Gene.* 2nd edn. Oxford: Oxford University Press.

Depew, D.J. and Weber, B.H. (1995). *Darwinism Evolving.* Cambridge, MA: MIT Press.

Edgar, L. (1992). Embryogenesis in *Caenorhabditis elegans*. In V.E.A. Russo, S. Brody, D. Cove, and S. Ottolenghi (eds) *Development: The Molecular Genetic Approach.* Berlin: Springer-Verlag.

Eldredge, N. (1989). *Macroevolutionary Dynamics*. New York: McGraw Hill.

Finkelstein, R. and Perrimon, N. (1991). The molecular basis of head development in *Drosophila melanogaster. Development*, 112, 899–912.

Ford, E.B. (1975). *Ecological Genetics*. London: Chapman and Hall.

Fraser, A.M. (1988). *Information and Entropy in Strange Attractors.* PhD dissertation, University of Texas at Austin.

Fraser, S.E. and Bronner-Fraser, M. (1991). Migrating neural crest cells in the trunk of the avian embryo are multipotent. *Development*, 112, 913–920.

Goodwin, B. (1985). Constructional biology. In G. Butterworth, J. Rutkowska, and M. Scaife (eds) *Evolution and Developmental Theory.* Brighton: Harvester Wheatsheaf.

Goodwin, B. (1988). Morphogenesis and heredity. In M.-W. Ho and S.W. Fox (eds) *Evolutionary Processes and Metaphors.* Chichester: Wiley.

Gottlieb, G. (1991). Experiential development of behavioural development: theory. *Developmental Psychology*, 27, 4–13.

Grobstein, C. (1978). *The Strategy of Life*. San Francisco: Freeman.

Grunstein, M. (1992). Histones as regulators of genes. *Scientific American*, 267, 40–47.

Ho, M.-W. and Fox, S.W. (eds) (1988). *Evolutionary Processes and Metaphors.* Chichester: Wiley.

Jacob, F. and Monod, J. (1961). Genetic regulatory mechanisms for the synthesis of proteins. *Journal of Molecular Biology*, 3, 318–356.

Keil. F. (1988). On the structure-dependent nature of stages of cognitive development. In K. Richardson and S. Sheldon (eds) *Cognitive development to adolescence.* Hove: Erlbaum.

Landman, O.E. (1991). The inheritance of acquired characteristics. *Annual Review of Genetics*, 25, 1–20.

Lawrence, P.A. (1992). *The Making of a Fly: The Genetics of Animal Design.* Oxford: Blackwell.

Levins, R. (1968). *Evolution in Changing Environments.* Princeton, NJ: Princeton University Press.

Lewontin, R.C. (1974). *The Genetic Basis of Evolutionary Change.* New York: Columbia University Press.

Mayr, E. (1970). *Population, Species and Evolution.* Cambridge, MA: Belknap Press.

Morisato, D. and Anderson, K.V. (1995). Signalling pathways that establish the dorsal-ventral pattern of the *Drosophila* embryo. *Annual Review of Genetics*, 29, 371–399.

Moran, N.A. (1992). The evolutionary maintenance of alternative phenotypes. *American Naturalist*, 139, 971–989.

Nijhout, H.F. (1990). Metaphors and the role of genes in development. *BioEssays*, 12, 441–446.

Oyama, S. (1985). *The Ontogeny of Information.* Cambridge: Cambridge University Press.

Philipson, L. and Sorrentino, V. (1992). Growth control in animal cells. In V.E.A. Russo, S. Brody, D. Cove, and S. Ottolenghi (eds) *Development: The Molecular Genetic Approach.* Berlin: Springer-Verlag.

Piaget, J. (1971). *Structuralism.* London: Routledge & Kegan Paul.

Plotkin, H.C (1988). The evolution of closed and open programmes of development. In D.R. Garrard and J.D. Feldman (eds) *Development in the Neocortex.* Cambridge: Cambridge University Press.

Plotkin, H.C. (1994). *The Nature of Knowledge.* London: Penguin.

Plotkin, H.C. and Odling-Smee, F.J. (1979). Learning, change and evolution: an inquiry into the teleonomy of learning. *Advances in the Study of Behavior*, 10, 1–42.

Powers, D.A., Lauerman, T., Crawford, D., and DiMichele, L. (1994). Genetic mechanisms for adapting to a changing environment. *Annual Review of Genetics*, 25, 629–659.

Purves, D. and Lichtman, J.W. (1985). *Principles of Neural Development.* Sunderland, MA: Sinauer.

Rendel, J.M. (1967). *Canalisation and Gene Control.* London: Logos Press.

Richardson, K. (1992). Covariation analysis of knowledge representation: some developmental studies. *Journal of Experimental Child Psychology*, 53, 129–150.

Richardson, K. and Webster, D.W. (1996a). Analogical reasoning in context: a research note. *British Journal of Educational Psychology*, 66, 23–32.

Richardson, K. and Webster, D.W. (1996b). Recognition of objects from point-light stimuli: evidence of covariation structures in conceptual representation. *British Journal of Psychology*, 87, 1–26.

Roff, D.A. (1992). *The Evolution of Life Histories*. London: Chapman and Hall.
Rollo, D.C. (1995). *Phenotypes: Their Epigenetics, Ecology and Evolution*. London: Chapman and Hall.
Rose, S. (1981). From causations to translations: what biochemistry can contribute to the study of behaviour. In P.P.G. Bateson and P.H. Klopfer (eds) *Perspectives in Ethology*, Vol. 4. New York: Plenum Press.
Russo, V.E.A., Brady, S., Cove, D., and Ottolenghi, S. (1992). *Development: The Molecular Genetic Approach*. Berlin: Springer-Verlag.
Scarr, S. (1992). Developmental theories for the 1990's: development and individual differences. *Child Development*, 63, 1–19.
Scharloo, W. (1989). Developmental and physiological aspects of reaction norms. *BioScience*, 39, 465–472.
Slack, J.M.W. (1991). From egg to embryo: regional specification in early development. Cambridge: Cambridge University Press.
Slobodkin, L.B. and Rapoport, A. (1974). An optimal strategy of evolution. *The Quarterly Review of Biology*, 49, 181–199.
Stearns, S.C. (1989). The evolutionary significance of phenotypic plasticity. *BioScience*, 39, 436–447.
Thelen, E. (1995). Time-scale dynamics and the development of an embodied cognition. In R.F. Port and T. van Gelder (eds) *Mind as Motion: Explorations in the Dynamics of Cognition*. Cambridge, MA: MIT Press.
Thelen, E. and Smith, L.B. (1994). *A Dynamic Systems Approach to the Development of Cognition and Action*. Cambridge, MA: MIT Press.
Waddington, C.H. (1940). *Organizers and Genes*. Cambridge: Cambridge University Press.
Waddington, C.H. (1957). *The Strategy of the Genes*. London: Allen & Unwin.
Waddington, C.H. (1975). *The Evolution of an Evolutionist*. Edinburgh: Edinburgh University Press.
Wills, C. (1991). *Exons, Introns and Talking Genes*. New York: Basic Books.

3

DEVELOPMENTAL SYSTEMS

Introduction

In the previous chapter I tried to show how the standard genetic assumptions underlying the current framework of nature–nurture debates in psychology are unlikely to be valid. This conclusion was reached simply from a brief consideration of genomic regulations, which appear to have evolved to deal with environmental perturbations that could easily throw direct expression from genes to characters 'off track'. These appear to work by, as it were, 'opening up' structural gene expression in a more responsive and flexible deployment. The enormous significance of this arrangement is highlighted by the fact that *up to 90 per cent of genes, and 95 per cent of the DNA associated with any particular gene*, even in fruitflies, may be regulatory in function.

Here, I want to continue this scrutiny by showing how the evolved system of genomic regulations is only the first of a number of regulatory levels to be found in organisms living in complex environments. Just as the expressions of structural genes have become nested in a complex of genomic regulations, so these have become nested in other systems of regulations, at additional hierarchical levels. As we shall see, development, instead of being merely the process of assembly of a predetermined phenotype, *has itself become the chief means of adaptation* in changing environments. In other words, I try to show how, instead of the organism being made up of predetermined phenotypes, including cognition, *systems of development* have rather become the key 'phenotypes' of organisms, each creating the abilities to deal with complex and complexly changing environments. This view will further contrast with the assumptions of cognitive ability as a fixed character, genetically designed to 'fit' stable recurrent situations, and individually varying accidentally with underlying genes.

Complex environments

Of course, psychologists tend to speak of the influence of the environment in at least three senses in considering the structure and variation in cognitive abilities. The first is the familiar one of the environment providing certain basic requirements (nutrients, oxygen, exercise, experience, etc.) that facilitate or attentuate the common expression of genetic potential – i.e. the 'environment of development'. The other (more recently introduced by evolutionarily minded psychologists) refers to the environmental selection pressures in our evolutionary past which, it is now thought, have shaped, and currently constrain, our cognitive functions: i.e. the 'environment of evolution' (see e.g. contributions in Barkow, Cosmides, and Tooby 1992). Finally, psychologists often refer to the variable aspects of environments that may be related to individual differences in cognitive ability – the 'environment of difference'.

The problem has been that almost all attempts to describe these environments have done so by stripping out the dynamic structure within them that has been the key to their significance, and reducing them to simple 'factors' or stable features. Yet without such an appreciation of complexity we can have little appreciation of 'higher' regulations. As Von Neumann (1966; cf. Bozdogan 1990) explained, the concept of complexity is '*a prerequisite to the understanding of learning and evolutionary processes*' (Von Neumann's emphasis). This neglect in understanding the environment of cognitive abilities is in stark contrast to the huge effort that has gone into proving their 'genetic basis'. The overwhelming tendency has been to reduce all environments to isolated 'good' or 'bad' factors (as nutrition is to the development of height), or stable structures to be adapted to.

In consequence of such environmental reductionism we still have difficulty understanding, for example, why species, in the course of evolution, have tended to become more complex, in the sense of more 'organised'. Such an extravagant outcome could hardly be the consequence of adaptation to fixed environmental contingencies, however complex they may be. In puzzling over this Darwin (1859: 11) pointed out that 'under very simple conditions of life a high organisation would be of no service', and may even be a disservice. We now know that there are other aspects of environments to which organisms have to adapt that cannot be described in terms of frequencies of 'good' or 'bad' factors. Rather, they have to do with the dynamics of change and variability – dynamics with which a purely genetic level of regulation would be incapable of coping. It is the evolution of higher forms of organisation or regulation in the context of such dynamics that this chapter is about.

Psychologists have only recently begun to be aware that the environment has structure, too, and that it is crucial to understand the role of that structure in cognitive development (Bronfenbrenner 1989; Wachs 1992). In what follows I want to show that increasingly higher levels of adaptive regulation, 'above', but incorporating and extending, genomic regulations, have evolved as species have had to contend with environments that are, first, changing more rapidly than can be dealt with by genetic selection (Plotkin and Odling-Smee 1979), and, second, more structured, in the sense that the predictability within them has to be found at increasingly deeper and/or shifting 'depths'. Such regulations depend upon information that has to be induced *de novo*, and cannot be already available in, or 'constrained' by, any one level alone. This, I shall argue, is what is at the root of cognitive regulations and human socio-cognitive abilities.

Developmental plasticity

Where the environment is consistently changing in a certain direction (for example, in supply of certain nutrients, temperature, aridity, and so on), as it has done over millennia, then new genetic resources have to be available, called upon, and 'selected'. This is the standard model of dealing with environmental change by 'genetic selection' of random mutations and reassortments. So long as change is directional and slow, relative to the reproductive generation time, such transgenerational regulations are adequate for maintaining life.

In other cases, though, environments may change 'directionally', but too rapidly to be adaptable to by genetic variation and selection – i.e. over periods shorter than a generation time. Or changes may even reverse upon themselves (Plotkin and Odling-Smee 1979). This may be because the environment is actually changing spontaneously, or because moving animals have moved to another 'patch'; because animals have changed their environments by their own activities; or combinations of these. Either way, the environment experienced by offspring may be different from that enjoyed by parents. How can adaptation by genetic selection cope with such changes? It clearly cannot, and new systems of adaptation are needed, over and above that offered by genetic variation and selection. If a species had nothing but selection from genetic variants to rely upon in such circumstances, it would soon face its demise.

The solution that has evolved is one in which the genomic regulations described in the previous chapter have been 'opened up'; instead of the developing system 'preferring' a definite end-point, the latter remains unspecified until environmental variables have been 'input' through current experience. This kind of epigenesis has been well described in animals, and is said to display *developmental plasticity* (changing in the course of development). It has also been called 'variable epigenesis', in contrast with the canalised epigenesis already discussed in Chapter 2.

Often, developmental plasticity results in quite different *types* of phenotype, even in the absence of genetic variation. For example, one species of barnacle reacts to the presence of predatory snails in its environment by developing a 'bent' form that is more resistant to predation compared with the more typical flat form (Stearns 1989; see Figure 3.1). Piaget (e.g. 1980) was struck by the way that water snails developing in pond habitats had elongated shells, whereas the same species developing in more turbulent lake conditions had more compact shells (Figure 3.1). As Dodson (1989) notes, some of these morphological changes – involving no changes in genetic constitution – are so great that they were once described as diagnostic of different species.

An even clearer example is, perhaps, that of snails which are predated upon by crabs on certain seashores. Although increasing the thickness of the shell offers protection, it is inevitably energy-costly and material-costly, and would be wasted if the snails end up on a crab-free habitat, as is often the case. This problem is compounded by the fact that such invertebrates disperse their young very widely, so cannot predict in advance where they will end up. The problem is solved by a developmental plasticity for thick shells or thin shells, which, in turn, appears to be triggered by the presence of some substance produced by crabs in the neighbourhood. 'It

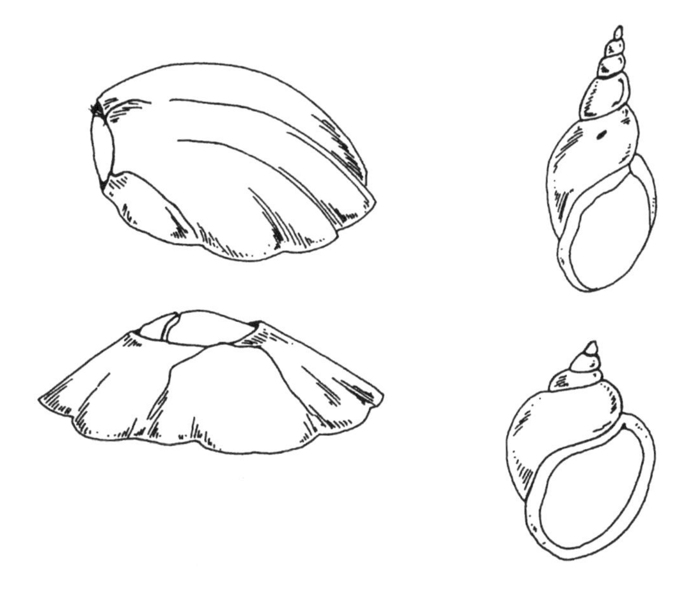

Figure 3.1 When a predatory snail is present in the environment of a species of barnacle it develops a resistent 'bent' form (top-left) instead of the normal form (bottom-left) (after Lively 1986). Water snails develop flatter shells in the more turbulent lake condition (bottom-right) compared with the form developing in pond conditions (top-right) (after Piaget 1980).

makes sense to leave the choice of shell morphology, and with it an entire trajectory of growth and reproduction, until after the adult habitat has been reached' (Stearns 1989: 442).

In other cases there appears to be more or less continuous variation. In mice, the tail, which is naked, acts as a heat radiator. Mice developing in warm temperatures appear to grow longer tails than genetically identical counterparts developing in cooler temperatures, probably because the extra length allows more efficient cooling (Waddington 1960; Johnston and Gottlieb (1990) review a wide range of cases of such temperature effects). Cichlid fishes, prominent in some African lakes, also show plastic differentiation in size of mouth parts in response to variation in shape and form of diet (Meyer 1987).

Remarkable degrees of developmental plasticity have been shown in early brain development in mammals. In one set of studies (Blakemore and van Sluyters 1975) cats reared in conditions in which they were exposed to lines of only one orientation developed a preponderance of nerve cells in the visual regions of the brain that were tuned to that *particular* orientation (as opposed to the full range normally developed). This suggests that the microcircuitry of the cortex (the most recently evolved part of the brain in mammals) is 'wired up' only in a rudimentary fashion at birth, with functional development dependent on actual experience (for review see Greenough, Black, and Wallace 1993).

Developmental plasticity in the brain is demonstrated spectacularly in experiments by Sur and his colleagues (see Sur 1993). They rerouted visual connections from the eye in newborn ferrets away from their usual destination in the visual cortex to what usually develops as the *auditory* cortex. What has long been recognised as the 'auditory cortex' developed, in these animals, into a *visual* cortex instead. This is evidence that, rather than being predetermined by a genetic 'code' for cortical functions, these functions emerge from other regulations operating at critical times *in the course of* development. Purves (1994: 67) has provided other examples of the induction of nerve cell connections, not by autonomous 'expression' from within the cells, but by their ultimate targets: 'In this manner, information that initially derives from neural targets can affect the connectivity of an entire pathway' (see also contributions in McGaugh, Bermudez-Rattoni, and Prado-Alcala 1995).

Nearly all our knowledge about developmental plasticity, discussed so far, has been about fairly simple morphological characters. This is not surprising. More complex systems of, say, physiological or behavioural functions are far more difficult to observe, analyse, and describe. None the less, it is clear that some epigenetic 'switches' occur among alternative physiological and behavioural systems, too. One example is the caste differentiation in bees and ants, which changes behaviour and physiology as well as anatomy. Another is the way that locusts develop different physiologically and behaviourally distinct 'morphs' in response to current population densities. These obviously reflect alternative preferences in a developing system.

A higher level of regulation

Developmental plasticity suggests a level of developmental regulations in which genomic regulations have become further embedded, yielding a higher level of 'ordering principles', as Goodwin (1988: 151) calls them. In each case different external conditions have shifted an interactive system from one stable state to another, passing through different developmental trajectories on the way and 'calling up' different gene products from the same or different alleles.

Although the existence of such regulation seems clear, more than general descriptions have remained obscure, except in very simple cases. One such example is that of a developmental switch in the gut bacterium *E.coli*. This can developmentally alter its membrane proteins to admit certain small molecules into its cells which would otherwise not be utilised. The regulatory switch consists of the effects of osmotic changes (changes in the fluid 'pressures' on either side of a membrane) on a two-step recruitment of gene expression. 'The first step changes a receptor molecule in the membrane. This change activates a control molecule in the cell that binds directly to the bacterial DNA and thus alters the expression of the genes that code for membrane proteins' (Stearns 1989: 441).

Whatever the detailed mechanism, adaptability of this sort clearly cannot be achieved by the independent expression of independent genes. Nor can they be achieved by 'canalised' genomic mechanisms alone. Rather, there is a 'higher' regulatory system at work. 'The epigenetic system, as we have noted, is highly integrated: each stage has its own system of regulations, and each is bound to the levels before and after by a complex of interregulations' (Piaget 1980: 66). In such a hierarchy of inter-regulations, lower systems become incorporated and 'opened up' in the still more creative process of development that occurs in the interactions *between* levels. It has also been shown how the resilience in such systems can accommodate or mask the expression of variable genes and new mutations (Rollo 1995).

Although we commonly speak of the genotype as 'creating' its appropriate phenotype through epigenetic processes, we now see that it is at least as legitimate to speak of the developmental process 'creating' its own appropriate genotype (Oyama 1985: 49). Figure 3.2 suggests how this entails a new level of 'information gain' distinct from the information in the genes (Plotkin and Odling-Smee 1979).

Unfortunately, although the ideas of 'epigenesis' and 'interaction' are

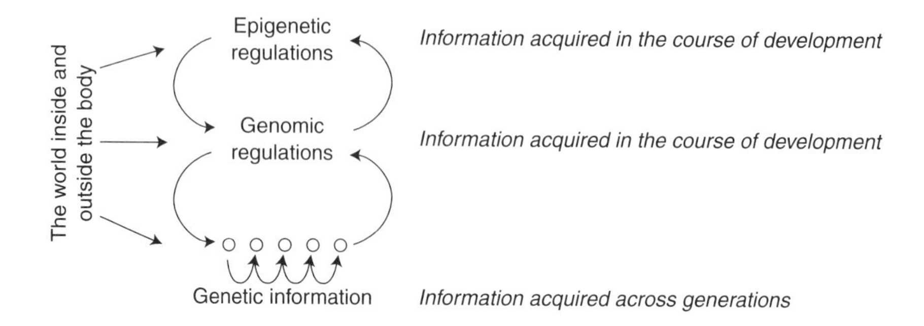

Figure 3.2 The epigenetic system guides development on the basis of information acquired in the course of development.

widely applauded, there has been a tendency simply to force them into the traditional framework of assumptions described in Chapter 1, leaving that framework essentially unchanged. For example, in attempting to partition genetic and environment variance underlying IQ, researchers who regularly subscribe to the idea of interaction continue to 'fit' statistical models based on purely additive assumptions (see Chapter 5).

Oyama (1985) indicates how terms like 'epigenesis' and 'interaction' have become like shibboleths to developmental psychologists, although they are often used imprecisely. All too often 'epigenetic' development is portrayed merely as an environmental attenuation or enhancement of a distinct 'potential' predetermined in the genes. Expressions like 'the interaction between the genetic and the social', or 'gene–environment interactions' are regularly used without details of such interaction ever being made clear. Instead, the picture is painted of a more or less harmonious amalgamation of randomly varying genes and randomly varying social or other environments partially enhancing or suppressing the effects of the other, resulting in a kind of accidental – and more or less 'good' – resolution in the phenotype. Such an anarchic picture is remote from the constructive reality seen in the examples above.

The point is that an organism in possession of epigenetic regulations becomes, as Goodwin (1988; 1991) explained, not a passive victim of either its genes or its environments, with fixed potentials, but an active source of *new* potentials. These arise, as explained above, in the interaction between levels. Indeed, some experiments have revealed an astonishing degree of 'hidden' potential in the epigenetic system. For instance, in evolution from reptiles, birds lost their teeth, and do not exhibit a gallery of teeth like mammals. This suggests that they have, therefore, no 'genes for teeth', and that it is surely impossible for them to *have* such teeth. Yet Kollar and Fisher (1980; cf. Gottlieb 1991) have shown that under certain conditions birds *can* develop a mammalian-type dentition.

Developmental plasticity as a lifelong process

So far we have seen models of genetic determination and found them to be inadequate for adaptive (or rather adaptable) characters in rapidly changing environments. Epigenetic regulations overcome this inadequacy by opening up genomic regulations previously closed around specific endpoints of morphological structure. But, as discussed so far, epigenetic regulations would appear to have a serious limitation, namely that they are concerned with 'one-shot' alterations of the phenotype. When fully formed organs or functions have appeared, this adaptability ceases. Once a snail has developed a pointed as opposed to a flattened shell it cannot change back. Once a barnacle has developed a bent as opposed to a straight form (Figure 3.1, p. 69), this cannot be reversed or altered or extended. Once a set of cognitive abilities has developed in childhood, we are stuck with them for life.

Yet many conditions crucial to survival may *continue* to fluctuate and change throughout the life of an organism. As already mentioned several times, this is particularly likely with animals that move around and/or change their environments through their own activities. What is required for successful living in such conditions are characters that can also continue to change *throughout the life of the individual.* There is abundant evidence for the existence of such continuously adaptable characters.

Clearly, the simplest lifelong environmental changes are those which are cyclical or oscillate about a predictable mean, or which 'switch' from one set of conditions to another across a threshold. Examples of adaptations to the former are the numerous physiological homeostatic processes such as temperature control. These involve sensitive interactions between numerous metabolites, cells, and organs, often involving cells themselves oscillating between one state and another. 'The cooperation and interaction of the organs establishes the organisms. In fact this is what is called physiology' (Nederbragt 1997: 151).

A more visible example is the fluctuating colour of the chameleon. Whereas the skin pattern of most reptiles is fixed at the end of development, that of the chameleon retains a lifelong changeability. Seasonal coat and/or colouration changes in a number of species are other examples of lifelong phenotypic plasticity. Sometimes, this plasticity can involve gross changes, as can be seen routinely in certain coral-reef fish:

> These fish live in spatially well-defined social groups in which there are many females and few males. When a male dies or is

otherwise removed from the group, one of the females initiates a sex reversal over a period of about two days, in which she develops the coloration, behaviour, and gonadal physiology and anatomy of a fully functioning male.

(Gottlieb 1991: 9, citing a study by Shapiro 1981)

Various 'inducible' phenotypes, responding to environmental signals across the lifespan, are now known. In some social insects, 'soldier' castes may be induced by collateral changes in their prey, such as the appearance of more defensive phenotypes in aphids. These changes are, as Rollo (1995: 202) notes, akin to rabbits changing to porcupines because wolves have been smelled in the woods!

But these are all cases of alternating 'switch-like' processes responding to a recurrent pattern of change, and are thus non-developmental. Lifelong environmental changes, especially those wrought by organisms themselves, can be both more rapid and much more complex than this. If the environment is *continuously* unpredictable, is not simply fluctuating in a given range, yet can provide novel situations, changing, perhaps, over periods of minutes or hours, then simple physiological switches will not do and more complex regulatory mechanisms have evolved. One example is the immune system: entirely novel antibodies may be produced throughout the life of the individual so that an individual 'profile' of antibodies continually develops according to the individual life history (Plotkin and Odling-Smee 1979).

More generally, though, in tracking lifelong, *rapid* environmental changes, behaviour – and, more importantly, regulations governing it – comes into its own. Behaviour, in fact, probably emerged as an instrument of adaptability very early in the course of animal evolution. It is changes of behaviour, and their regulations, above all else, that have determined the courses of evolution of animals over the last hundreds of millions of years (Mayr 1970; Bateson 1988). Yet behaviour and its regulatory change (cognition and learning) are still poorly understood.

One reason for this may be that behavioural biologists and psychologists have tended to focus on relatively 'closed' or 'instinctive' behaviours, such as mating, nesting, and territorial routines. Related to this has been an abiding tendency to think of behaviour in terms of narrow 'adaptive' responses to well-defined 'stimuli', as per the adaptationist paradigm. These 'closed behaviours' may present a uniformity and repetitiveness convenient for scientific observation and analyses, but they are untypical of evolved behaviours and their regulation (Mayr 1974). Although they have constituted, over the last fifty years or so, an exciting biology of behaviour, they miss a crucial stratum of higher behavioural regulations that requires a *psychology* of behaviour for proper understanding.

This is where we need to think a little more about environmental structure. Obviously, no organism could survive where predictability (in food, predator behaviour, temperature extremes, and so on) is completely absent. But a superficial absence of predictability may belie its availability at some deeper level, because of environmental structure. Adaptation to *changeable* environments is likely to rely precisely on that deeper predictability within it, as furnished by environmental structure.

There is environmental structure when a crucial variable such as food availability is too unpredictable in itself for an organism to adapt to, yet can be predicted from other variables with which it covaries. Consider, for example, a blue-tit which experiences nuts sometimes in location A and sometimes in location B, but does not have the energy to reach both locations in a single day. Even by randomly setting its targets, the bird will, at best, be eating irregularly. Now assume that the location varies with some other variable, such as the weather (A when it is fine, B when it is wet). If the bird has a behavioural regulatory system sensitive to such environmental covariation it will feed much better.

But that covariation may be also be weak – giving better, but still insufficient, predictability about food location. Now consider how that covariation may itself be conditioned by another variable such as temperature, so that the food is most likely to be in location A when it is wet and warm, but location B when it is dry and warm; A when it is dry and cool, but B when it is wet and cool. A bird able to capture this 'deeper' covariation structure will render food availability far more predictable than may seem possible from casual inspection (see Figure 3.3).

Such informational structure may, of course, be much more complex, including many more values/variables and hierarchical levels of conditioning or interaction (Figure 3.3). A description of a system in terms of the interactions of its components or variables is a natural measure of its 'complexity' (van Embden 1971; Bozdogan 1990). Although 'surface' variations of single factors, or even bivariate associations, may appear too capricious for adaptation to them to be possible, deeper interactions among them, where they exist, can help make complex environments predictable. Obviously, an organism that can 'tune in' to such structure may thus flourish where others would not.

Note that many such contingent structures may be stable over long periods of time (i.e. generations), in which case adaptation to them can be regulated by evolved switch mechanisms at an epigenetic or genomic level, without the involvement of cognitive mechanisms at all. One of the simplest examples of 'tuning' to deeper informational structure can be found in the contingencies that govern behavioural adaptations to seasonal changes (Colwell 1974). Thus migration, mating, and so on, become adaptive to a varying environmental parameter (e.g. food resources) through the *latter*'s covariation with an intermediary (temperature or photoperiodicity). Changes in light/temperature thus signal food resources

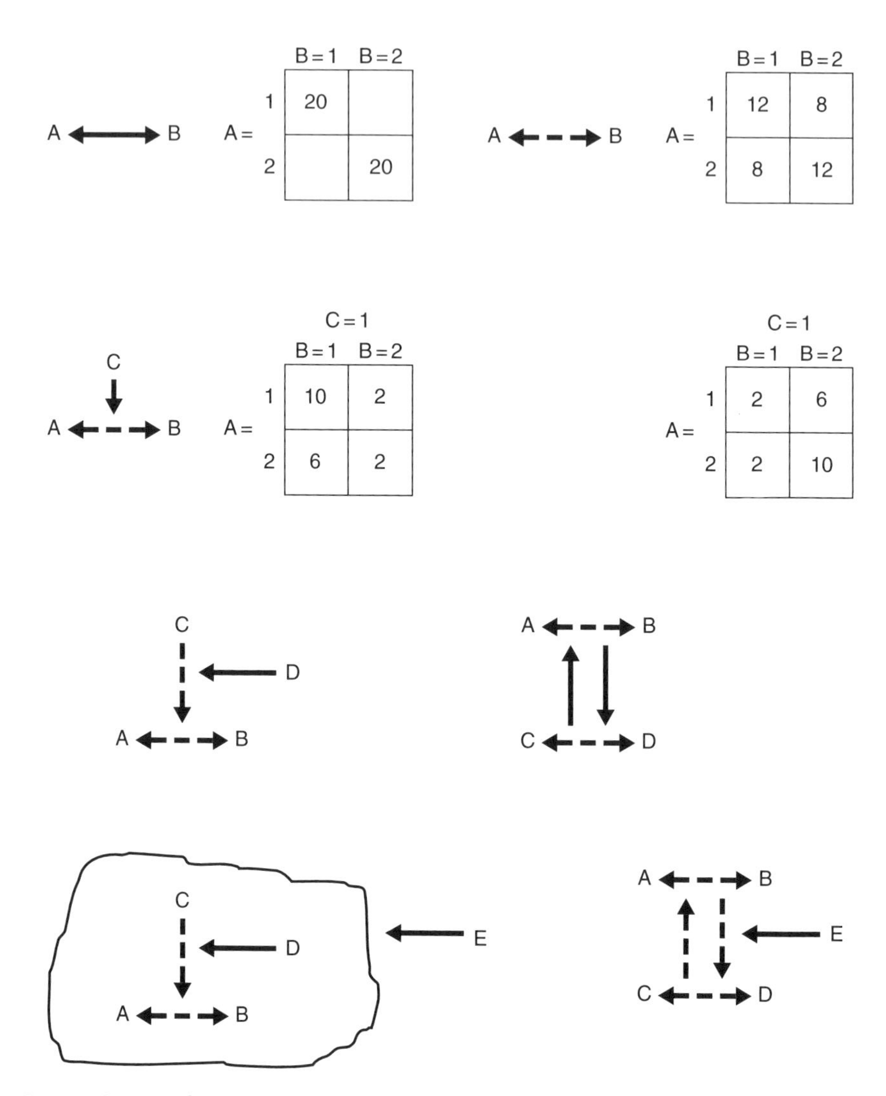

Figure 3.3 (a) Complex environments may have little immediate predictability; but various kinds of informational structure can yield it. The simplest structure is a perfect covariation between two variables (A and B), making the values of one predictable from values of the other (top-left, which also indicates expected frequencies of combinations in a contingency table). In real life, though, such covariation is likely to be less than perfect (top-right). Predictability may again be improved, though, if the weak covariation is embedded in, or conditioned, by a third variable (second row). The rest of the figure indicates how, even for a world made up of binary-valued variables, a wide range of environmental structures is possible.

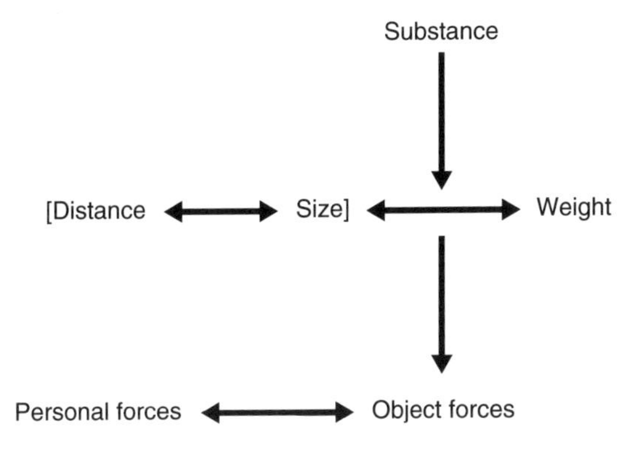

Figure 3.3 (b) A fragment of a nested hierarchy of covariations involved in grasping an object.

at some future time or in another place, and a corresponding behaviour change *now* can anticipate them just when they are most needed for the new broods ahead. Clearly, this covariation needs to be captured or 'represented' in some internal regulatory system – in this case an evolved system of hormonal and neural regulations.

A number of such 'contingent' adaptations have been described at an epigenetic level. Pittendrigh (1958) dicusses how covariation between day/light transitions and atmospheric moisture permits the emergence of a Drosophila fly from the puparium at dawn, thus avoiding dessication that would follow later in the day. Johnston (1982) discusses how the fact that day length predicts later daily temperature in the environment of one species of butterfly allows the developmental system to lay down wing pigmentation (a thermoregulatory mechanism) in advance in the larva, but not in a related species, the environment of which contains no such covariation structure (cf. Oyama 1985: 135).

But, as indicated in Figure 3.3a, much more complex structures are possible. For example, the covariation between food and length of day may itself be contingent upon (conditioned by) latitude, altitude, or other physical conditions. There is now a new 'depth' to the contingency. Some of these deeper relations can theoretically be extremely complex, not only in the sense of increasing 'depth', but also because the covariations at all or any depth may be non-linear. In consequence, extremely variable and complex patterns of covariation structures can be found in nature.

As already mentioned, adaptation to such deeper contingencies at an epigenetic level is only possible if the contingencies themselves are sufficiently stable and reliable over long periods of time. Now imagine what happens if a species is faced with important aspects of the environment not only lacking in 'direct' predictability, but also in which any deeper patterns of contingency useful for survival *are themselves changing more or less frequently within the lifetime of the individual.* This may well be so in the case of our blue-tit mentioned earlier if the person, whose kind habits the tit has assimilated, works shifts and ends up putting food out at different times of the day. As another example, consider a monkey that feeds on fruit in the upper canopy of a rainforest (having discovered a covariation between elevation and food abundance). The monkey also learns that nearer the forest margins the fruit lies lower down (a height/abundance covariation conditioned by location). But those marginal locations, at certain times of the year, are also more frequently scanned by monkey eagles (the conditioning itself conditioned by a correlation between predator presence and season). This covariation structure – which makes the environment predictable through its 'deep' contingencies – may change completely when the troop moves to another forest because of fruit depletion (i.e. its own activity).

The seemingly simple task of grasping and handling objects (by mouth or hand) also conceals the complexity or depth of forces involved in the precision of predictability that arises. Such action requires, on the basis of past experience, a preparatory tension in an array of muscles which must exactly match an array of perceived forces in the object, such as shape, orientation, speed and direction of motion, size, and weight. But this covariation of action will be conditioned by covariation *between* some of these forces: for example, size and weight. And these covariations may may be conditioned by others at further depths. For instance, perceived size (and weight) will be conditioned by perceived distance; and the size–weight relation will be conditioned by perception of substance (fruit or rock). Figure 3.3b indicates just a fragment of such a hierarchy of forces.

All of these deeper covariations must have been represented in some form, on the basis of past experience, on a continually renewable basis, in order to make predictability possible. In the real world of evolved animals, rarely are two objects, or their dispositions with respect to the individual, exactly alike; so there is 'constant novelty'. It is to help us deal with novel futures on the basis of limited past experience that cognitive systems evolved. I hasten to add that even the nested covariation structures already described are crude parodies of the richness of the dynamic structures available in real environments, especially social ones (to which I turn in the next section).

If the world of experience is changing more or less rapidly in the course of a lifetime, dealing with it requires more than stable, 'built-in' models of contingent cues – i.e. 'innate' regulations and behaviour patterns. It requires a whole new layer of regulations that can capture those complex contingencies in some other system of representation on a continually renewable basis – and, of course, construct motor actions *de novo* to ensure the most appropriate response to the present situation. It is the need to capture this deeply conditioned information that seems to explain the mushrooming cerebral cortex, and with it the cognitive system, in the evolution of vertebrates.

The commonly used expression 'learning', to describe such adjustments is, of course, correct. However, because of psychologists' descriptions of learning in terms of simple associations, the term usually fails to capture the 'deeper' structures being described here. Likewise, knowledge, even in its rudimentary forms in non-human mammals, consists of more than a device for tracking increased *frequency* of change, as Plotkin (1994) suggests; rather, it is the informational structure through which environmental predictability is obtained that is crucial.

I have argued in several places (e.g. Richardson 1992; Richardson and Webster 1996) that such learning is only achieved by the lifelong developmental construction of 'nested covariation hierarchies' within the cognitive system reflecting those in natural experience, including those revealed by action upon the environment. This is the kind of information that cognitive systems evolved to deal with. Indeed, a number of neuroscientists have long argued that it is such nested covariations that form the 'language' of higher cerebral functions (e.g. Mackay 1986). This is the kind of informational structure which organisms had to be able to capture in order to make their lifelong changeable worlds predictable. Such 'deep' information forms something of a structural 'grammar' (knowledge) which also serves as the structure of reasoning (cognition) in rendering predictability from (superficially) chaotic experience. By means of such a grammar, organisms with cognitive systems can begin to make predictions about situations on the basis of highly fragmentary information (Richardson and Webster 1996), or from temporally or spatially remote positions (i.e before they have been directly encountered), and thus even 'rework' the realities of the world they have experienced and *change* them in advance of their needs.

Note that, just as epigenesis 'opens up' genomic regulations, delaying their commitment to one phenotypic form or another, so cognitive regulations 'open up' epigenetic regulations in an analogous sense. This is most clearly seen in the way that epigenetic processes are recruited in, for example, the plasticity of cortical cell connections *on a lifelong basis* (Weinberger 1995). Such plasticity is over and above the experience-dependent pruning of neural connections that forms the basic architecture of the brain in early development (e.g. Huttenlocher 1993).

Although we commonly refer to these lifelong changes, and their biochemical, epigenetic underpinnings somewhat loosely as 'learning' and 'memory' (for reviews see Rose 1992; McGaugh, Bermudez-Rattoni, and Prado-Alcala 1995), Singer's (1990) expression, 'experience-dependent self-organisation', may better reflect the complexity and depth of informational structure that is entailed. In any case, it seems clear that cognitive regulations are not simply products of epigenetic development: they represent an entirely *new* level of information 'gain' (c.f. Plotkin and Odling-Smee 1979), superseding and incorporating epigenetic regulations, as indicated in Figure 3.4.

As we shall see shortly, there is still a further level of cognitive regulation among humans which I turn to in the next section. But already it should be clear how this view of nested regulations contrasts with the simple model almost universally subsumed (implicitly or explicitly) in nature–nurture debates about cognitive ability. The latter simply depicts cognitive ability as a 'quantitative' character, the form and variation of which can be related to independent genes and environments exactly as can physical characters like height. That is, all adaptive functions are founded on a singular causal model, which is that of an addition of genetic and environmental forces (the 'summary of forces' model shown in Chapter 1, Figure 1.8, p. 25).

When we seriously consider the complexity of events to which cognitive characters are an adaptive (or rather adaptable) strategy we see that cognitive regulations are on a completely new plane, and that the application of simple adaptationist or quantitative models to them seems singularly inappropriate. An analogy might be attempts to sum up the proceedings of a dinner party in terms of ingredients and recipes involved in the food preparation before the guests have even arrived.

I stress again how new potentials (this time for knowledge and reasoning) arise in the interactions between levels, and are not prefigured in

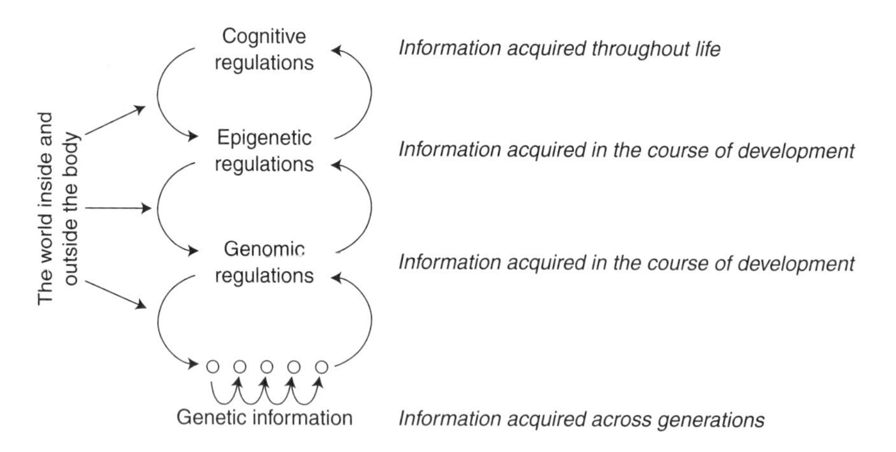

Figure 3.4 Cognitive regulations achieve adaptability in lifelong changeable circumstances. In the process they 'open up' and utilise epigenetic regulations throughout life.

genetic, genomic, or epigenetic regulations alone. The main problem among psychologists has been that of *agreeing* about the nature of these cognitive regulations. As described in Chapter 1, nativists have insisted that it requires prior information in the genes laid down by natural selection. Another group of theorists, the Associationists – also stemming from Ancient Greece – have argued that the cognitive apparatus simply 'copies' associations experienced in the environment. This view neglects, though, the fact that predictable associations, especially those in the kinds of complex environments just described, may not consist of simple, immediately available correlations. There is currently much debate about whether the latest wave of associationist theorising – so-called 'connectionism' – overcomes this problem (see, e.g., Bates and Elman 1993).

The great strength of 'constructivist' theories such as that of Piaget (e.g. 1988) is that it emphasises the complex 'coordinations' in experience, revealed by action, which are *more* than direct copies from experience. Piaget's great weakness was that he attempted to describe the content of these coordinations by analogy with propositional logic (Johnson-Laird 1983). None the less, it is clear that Piaget's life work aimed to describe the developmental construction of an increasingly creative level of adaptive functioning autonomous of both genes and experienced environments. The most detailed description of this creativity is in the passage from sensorimotor to representational coordinations in the average two-year-old. The latter fosters a rapid, continual restructuring of reality as directly experienced, and thus of adaptive possibilities, without the haphazardness of prestructured forms or trials and errors:

> Henceforth there exists invention and no longer only discovery; there is, moreover, representation and no longer only sensorimotor groping. These two aspects of systematic intelligence are interdependent. To invent is to combine mental, that is to say, representative, schemata and, in order to become mental the sensorimotor schemata must be capable of intercombining in every way, that is to say, of being able to give rise to true invention.
>
> (Piaget 1936/1962: 341, quoted by Marti 1996: 62)

In this process, the coordinations of the sensorimotor schemata become freed from immediate perceptual data and actual action, and can rapidly and autonomously range over more discrete predictions and *possible* actions. These anticipatory and exacting functions, occupying a wholly new time–space compared with epigenesis, constitute a new layer of regulation. Adaptive modifications are now achieved by extremely rapid and purposeful *cognitive* 'mutations' and 'reassortments'. These come to be seen immediately in the child in the form of deferred imitations, mental imagery, symbolic play, and language use. In each of these, instead of endogenously constrained sterotyped patterns, there is a new adaptability of thought and behaviour. With these developments come a new subjectivity and intentionality not foreseen on previous levels. In later stages of development, this inventiveness comes to be amplified many fold.

As Marti (1996: 62) notes, one consequence of this freer and rapid regulatory function is that 'it is less visible and frequently escapes the eye of the observer'. This descriptive difficulty may be one reason why human cognitive abilities so frequently get reduced to simple response parameters 'matching' stereotyped expectations (as in IQ tests and other forms of assessment). However, such an account is already anticipating another level of regulation in humans, and this is what I move on to in the next section.

Human socio-cognitive regulations

Cognitive systems evolved when predictability for survival only arises from 'patterns in depth', and such patterns change very quickly, or new ones are continually arising, throughout the lifetime of the individual. Although they already take us a long way from the simple expression of genes in Mendelian traits, along with isolated environmental factors, they are by no means the last layer of regulations, nor the greatest. So far we have been dealing with cognitive regulations and their development as isolated mechanisms locked within the individual, producing adaptations that are the sole 'property' of that individual. But another method of adaptability has evolved that goes beyond these.

Obviously, the deeper the covariation structures that a species can abstract to create predictability, the richer it is in terms of cognitive ability. Even judged at an individual level it would be clear to any visitor from outer space that humans are very special in this regard. Even everyday, moment-to-moment activitics entail the use of patterns of covariation at several depths, almost always involving what *other* people are doing. Take, for example, waiting for an unreliable bus (a mode of transport that is itself a prominent social 'tool', of which more below). You may realise that, for various social reasons, the bus is consistently late on Tuesdays and Thursdays and particularly late on Fridays. So that, although its arrival is variable, there is a pattern to the variability. Instead of perishing in the cold out of force of habit you can get used to the pattern by a corresponding pattern of behaviour. The pattern itself might change, however, when the schools are on holiday, so that the buses may even arrive early on some days. So your knowledge has increased and you can adjust to *changing* patterns as well. *There is no other species on earth that can exhibit what is, for humans, this extremely mundane kind of cognitive ability.*

Languages, games, practical skills such as driving, everyday social interactions, and even simple computations, appear to have this more abstract kind of structure, in which events at one level, because of their social embeddedness, are dependent upon events at another level, through, possibly, several layers of dependency (Richardson and Webster 1996). These do not constitute a 'universal physical law' governing the 'survival of the most stable' (Dawkins 1989), except in a highly abstract sense. Even very young children appear to be adept at acquiring and using such abstract social structures in their everyday reasoning (Dunn 1988).

How has this remarkable, and universal, cognitive ability come about in humans? Here I argue that it has arisen as a product of the evolution of another, even more spectacular level of developmental regulations, namely social regulations. Ordinarily, cognitive representations will take time to build up and will be inevitably limited by what the individual actually experiences in one-to-one relationship with their environment. In humans the multi-level system of regulations we have seen so far is itself embedded in a uniquely human, higher level – a social level – which has proved to be impressively potent.

The evolution of human social regulations

The most obvious physical correlate of human cognitive abilities is, of course, our enormous brain. The average human brain is three times that of our nearest biological cousins, even when account is taken of differences in body size. And the main changes have been in precisely those areas thought to be responsible for the processing of more abstract information structures (Figure 3.5). So it is naturally thought that our individual cognitive 'cleverness' simply arises from the size of our brains.

At the same time we have to remember that this enormous size comes at an equally telling cost. The brain is an energy-expensive organ. In adulthood, the brain consumes nearly a quarter of our energy use; in children this can be up to 50 per cent. The larger head it requires also complicates birth, and demands a long period of post-natal growth, with heavy investment of care from parents (Byrne 1995). Yet, once triggered, the expansion took place at an incredibly rapid rate, suggesting the evolutionary 'discovery' of a very great benefit. Accordingly, the human brain has been searched for special attributes which alone might account for these advances, and which in turn may be attributed to 'special' genes.

As striking as such changes are, however, it has become increasingly clear that our cognitive ability as a species is not based so much on raw processing capacity as on the *kinds* of information this 'new' brain needs to capture and process. For example, there is a rough correlation across

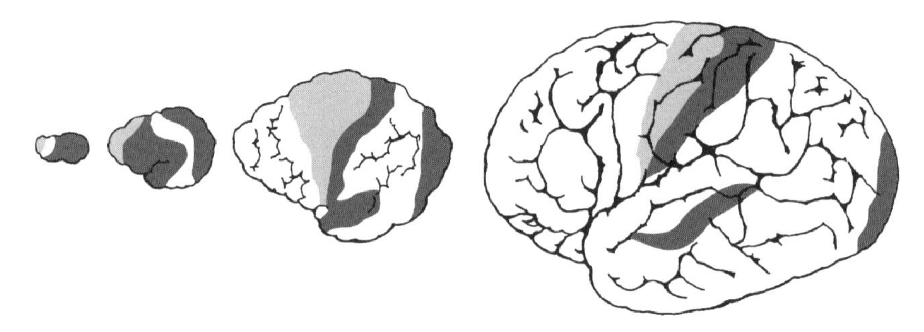

Figure 3.5 Drawings of the brains of four mammals showing the comparative sizes, including increased folding, of the 'new' areas concerned with adaptability. 'New' association cortex in white; 'older' areas shaded.

species between brain size and learning: as brains get bigger from fishes to apes (relative to body size), there is a corresponding increase in 'brain power' (Jerison 1973). But among humans, 'normal' and even exceptional intellectual achievement is found among people with brain sizes between $950cm^3$ and $1750cm^3$ (and there have even been normal cases well below the lower limit of this range).

Instead of crude size, therefore, we have to think in terms of more subtle functional requirements: we need to look more closely at what this enormous brain 'is for' in humans. 'The important difference between us and our nearest relatives might on balance have less to do with our brain's size, and more to do with how we use it' (Richards 1987: 143). In other words, our huge brains, including massively increased numbers of neurons and their interconnections, may be a *consequence*, rather than a primary cause, of the kinds of regulations it is called upon to make.

So how *do* we use it? What is this enormous brain for? Psychologists have examined human evolution for a closer understanding of *psychological* functions, and of how they develop. Although this is still a rather murky topic, dominated by often-exaggerated interpretation of small fossil fragments, living sites, food remains, and so on, millions of years old, it is also an exciting one. Here I can consider, only very briefly, three stages that have been broadly identified.

Walking on two legs (bipedalism)

Human ancestors evolved from tree-dwelling primates at least fifteen million years ago when climatic changes led to thinning of the forests. Little is known about the habits of these forebears until about four million years ago, when fossil remains and even a few footprints show that they were now bipedal (walking more or less upright on hind limbs).

Not much is known, though, about what advantages lead to the evolution of this new posture, but it was probably related to the new, sparsely wooded terrain of long grasses and leafy thickets. However,

> The bounteous benefits of the bipedal mode once adopted are abundantly clear: our hands are freed to carry, make tools and gesture, our forward facing heads atop a vertical spinal column maximise our range of vision, hands can add another sound channel to the vocal one by clapping, drumming etc.
>
> (Richards 1987: 154)

Another possible benefit was that it permitted the huge expansion of brain and cranium size which followed – one which would have put an impossible strain on the forward-supporting neck musculature of a quadruped. So it is thought that bipedalism was an important precursor to human cognitive abilities.

Tool manufacture

The freedom of the hands certainly allowed these creatures to make tools. There is abundant evidence of tool manufacture and usage by human-like (hominid) creatures two-and-a-half million years ago. Although rudimentary tool usage is found in many species (as when a thrush uses a stone as an anvil for smashing snail shells, or chimps use a stick to 'dip' for termites) early human tools appear to have reached new levels of sophistication.

Many theorists have speculated about the intellectual significance of this. Parker and Gibson (1979), for example, explain how manufacture of the kinds of tools discovered requires such intellectual skills as:

1 object–object coordination as when one article is used to produce, by hammering or slicing, a specific consequence in another;
2 discovery of new means–ends relations (e.g. conceiving of the distant consequence of a long chain of actions);
3 organising these in a hierarchical sense, as when two superficially disparate actions (e.g. slices in a stone tool) combine functionally in the context of an overall design;
4 abstract mental representation of events (e.g. of the novel task the action is intended to solve).

Here, again, though, pressures for improved tool-making may account in part for the emergence of the bigger brain in such creatures, but it cannot be the whole story. At this stage of evolution, these creatures still had brains only slightly bigger than those of modern apes. So it is thought that additional factors played a crucial part.

Social cooperation

Most theorists have stressed the importance of social relations for brain and psychological evolution and, in particular, the benefits of cooperative foraging, hunting, and defence. The excavation of archaeological sites has suggested that by two million years ago our ancestors were living in small social groups occupying temporary base-camps. From these sites, individuals or subgroups may have travelled each day, foraging for vegetation and either hunting or scavenging for meat which they butchered using crude tools. Food was then, perhaps, brought back to the camp and shared, much as is seen in some social animals (like African wild dogs) today.

Richards, quoting Lovejoy (1981), explains how this arrangement

> puts a premium on the ability to exchange information and to make arrangements regarding future movements of group members. It also increases the importance of regulating social relations among individuals. All these influences might be expected to favour the evolutionary development of an effective communication system, such as protolanguage, and of sharpened socio-intellectual capabilities.
>
> (Quoted in Richards 1987: 162)

From these early foundations, complexity of social life and increase in brain size seem to have evolved rapidly together. It seems that human brain functions evolved not simply to create individual 'cleverness', so much as to meet the needs of an increasingly successful social way of life. The complexity of social-technical life became the context of increased brain complexity which made even more complex social-technical life possible, and so on. This virtuous spiral proved to be spectacular. 'Modern' humans appeared only about 100,000 years ago and rapidly displaced all competitors such as the Neanderthals. Nearly all human change since has consisted of social and technological change on the basis of developmental systems that were then in place.

It is important to stress how the regulatory demands on the brain of operating jointly with other individuals are far more complex than those presented by the physical world alone (Humphrey 1976; Doise 1988). If you have any doubts about this consider moving a wardrobe downstairs with two other people. You need to integrate a multitude of stimuli, changing over fractions of seconds, continually adjust personal actions correspondingly, regulate personal feelings, and communicate rapidly and efficiently. There are few humans who cannot perform tasks like this, or, indeed, others far more complex; there are no other animals who can do it.

Increasingly complex social life also demanded communication. The crucial role of a uniquely structured language for mediating joint attention and joint action has been described by a number of psychologists (Vygotsky 1962; Bruner 1983). But such communication became important in another sense, too. If a member of a group can pass on the acquired structural representations of a class of problems to other group members, or to the young, then the behaviours of others are modified, and become 'pre-adapted' for those situations, with a minimum of actual experiences of them. The means by which such transmission of representations and behaviours (in sum, cultures) is passed on this way has recently become a very fertile field of research (Cole 1988; Rogoff and Chavajay 1995).

Thus, although there are disagreements and many other factors, most contemporary authors stress the importance of social behaviour in human evolution as the basis of our cognitive abilities. It is important to appreciate how this behaviour is both continuous with, yet different from, that of our primate ancestors. As Lancaster notes,

> crucial elements of the human pattern such as bipedalism, tool using and tool making, food sharing and cooperative hunting of meat do occur in non-human primates but only as relatively minor behaviour patterns and not as key adaptations. These elements come together in early humans coupled with a fundamental change in social and economic relations within the group.
>
> (Lancaster 1975: 78)

The same point has been put in different ways by many authors. 'The intense social interaction and regulation of social relationships required is . . . a more likely single cause for the origin of human intelligence than any other . . . demands on intelligence are greater in dealing with the social world than in tackling the external environment' (Richards 1987: 171; see also Byrne 1995). Or, as Dunn put it, from a developmental perspective:

> To become . . . a member of that complex world – children must develop powers of recognising and sharing emotional states, of interpreting and anticipating others' reactions, of understanding the relationships between others, of comprehending the sanctions, prohibitions and accepted practices of their world.
>
> (Dunn 1988: 5)

Humans appear to have evolved a still further method of 'information acquisition and storage' – an *interpersonal* as well as an *intrapersonal*

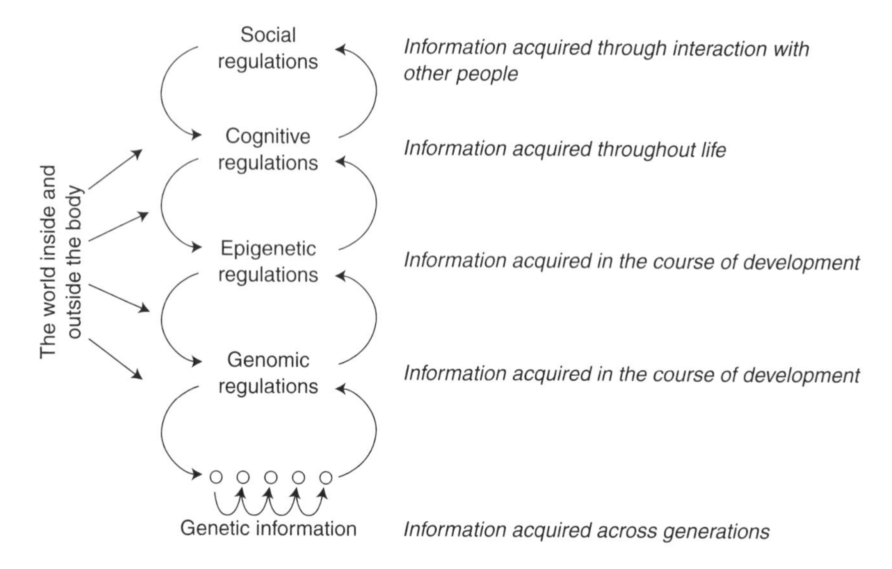

Figure 3.6 Social level of regulations superimposed over other levels of regulation nested within it.

one – and it is the regulation of such a system, at least partly, that our huge brains are for (Figure 3.6). It is not difficult to see how other evolved regulations are nested in it. Human knowledge, cognitions, needs, and motivations become contained and expressed within a social frame. Just as cognitive regulations 'opened up' epigenetic regulations (and these opened up genomic regulations, etc.) so social regulations have opened up cognitive regulations, permitting dazzling new potentials to be created perpetually in the interactions between these levels. I will have much more to say about this process, in what I call the cognition-culture complex, in Chapter 6.

Cultural tools

Clearly the cognition-as-self-organising-process (such as an internal representation), described in the previous section, must somehow be a common or shareable one. And it must be manifest in 'versions' that are shared at the same time as being changeable. Such shared models about the world, in connection with cooperative and technical processes through which it is experienced, constitute what is usually referred to as 'human culture'.

Humans attend to and act in the world as social conglomerates: their needs and aspirations are satisfied through social organisations, not individual adaptations. The regulations which these organisations comprise is known as human culture. At it strongest, this view means that individual development is absolutely dependent on the acquisition of culture:

> [the human] nervous system does not merely enable [us] to acquire culture, it positively demands that [we] do so if it is going to function at all. Rather than culture acting only to supplement, develop and extend organically based capacities . . . it would seem to be ingredient to those capacities themselves. A cultureless human being would probably turn out to be not an intrinsically talented though unfulfilled ape, but a wholly mindless and consequently unworkable monstrosity. Like the cabbage it so much resembles, the Homo sapiens brain, having arisen within the framework of human culture, would not be viable outside of it.
>
> (Geertz 1962: 723–724)

Survival in the world by cultural regulations, which embeds other levels of regulations, is far more adaptable and successful than those other regulations alone. Instead of adapting through the agency of evolved physical organs, humans have evolved the capacity for creating and recreating social-technical 'organs' as the means of existence, from the earliest stone tools and social organisations, to modern electronic and other devices, and complex local and international webs of cooperation. In addition to developing organic 'tools' of survival, like the eye and the hand, therefore, humans have the ability to acquire 'cultural tools' that are far more adaptable. As Donald put it,

> Our genes may be largely identical to those of a chimp or gorilla, but our cognitive architecture is not . . . humans are utterly different. Our minds function on several new representational planes, none of which are available to animals. We act in cognitive collectivities, in symbiosis with external memory systems. As we develop . . . we reconfigure our own mental architecture in nontrivial ways.
>
> (Donald 1991: 382)

The concept of cultural tools as a distinct level of regulatory activity has become very important in studies of cognitive abilities in recent years. Among them Vygotsky (1978) included all cooperatively organised activities: organised industrial production; various inventions for 'thought sharing' (e.g. number systems, language and writing systems); schemes for cooperative action (such as shared plans and explicit, scientific theories); a myriad social rules and principles (such as motoring rules); as well as 'hardware' tools, machines, and other technological devices. They incorporate regulation from the cognitive, epigenetic, and still lower levels into a new and higher level. As Vygotsky explained,

> By being included in the process of behaviour, the psychological tool alters the entire flow and structure of mental functions. It does this by determining the structure of a new instrumental act just as a technical tool alters the process of natural adaptation by determining the form of labour operation.
>
> (Vygotsky 1981: 160)

What the child or other novice acquires in the form of cultural tools are the nested covariations of joint or shared action, often focused around tangible instruments such as a knife or a computer. The Soviet psychologist, Leontiev (see Scaife 1985) illustrated this with the example of a baby feeding with a spoon. At first the infant handles a spoon as he or she would any natural object. But this handling becomes slowly, but radically, reorganised by adult intervention to conform to a more specific social use (see Figure 3.7). We only have to remember the diversity of eating habits and eating tools around the world to appreciate how this form of development is far from that of maturation of fixed behaviours defined in advance by a genetic code. Figure 3.7 shows four common cultural tools: feeding with an implement; a game; literature; a committee.

The point about these abilities is that they do not merely furnish reactions to complex events: the rich conceptual representations that social life develops allow us to know and cognise about the structure of the (socially mediated) world such as to anticipate and *make* change. Humans, particularly through the regulations called science and technology, adapt the world to themselves rather than vice versa. In the process, the constraints of other levels of regulation are overcome all the time. The constraints inherent in our joints limit the movements of our limbs. But these constraints are completely overcome through cultural tools. Humans do not have the epigenetic and other regulations that, in other species, have produced wings and flying – but we do fly thanks to the evolution of another, far more potent, level of regulation. The same thing can be said about other abilities such as moving underwater, underground, across deserts, and so on. Since their first appearance, modern humans have overcome every ecological barrier, survived in every available niche, and radically changed their own 'minds', often several times in the course of a lifetime.

Again, it is important to appreciate the dialectical 'method' whereby this unique developmental system operates in producing not only novel responses rapidly in day-to-day situations, but, on occasions, revolutionary changes in whole cultural patterns. This process is precisely one of interaction between the social level of regulation and that of the indivi-

Figure 3.7 Development as the acquisition of cultural tools.

dual cognitive level. It is through interactions in what I will call the 'cognition-culture complex' (see Chapter 6) that new potentials constantly arise.

As Vygotsky (1978) argued, cultural change, new practices, and new ideas arise because the relationship between individual and group is a dynamic, interactional one, not simply a handing over of a tradition. The child's or novice's mind is not a computer passively waiting to be programmed, nor that of an apprentice simply 'picking up' what the adult already knows and does. Rather, it already contains representations of countless other experiences, themselves at various levels of abstraction. These will often conflict with, and react to, patterns currently being internalised (Vygotsky used the word 'clash'), producing novel resolutions. 'The very essence of cultural development is in the collision of mature cultural forms of behaviour with the primitive forms that characterise the child's behaviour' (Vygotsky 1981: 151). In this way, the cultural order is, in a sense, reconstructed in each developing child, emerging in 'an individually differentiated fabric of meaning' (Brockmeier 1996: 133). This relationship between individual cognition and social forms allows the 'original ideas and creative contributions of individual minds' while explaining 'the achievement of social history' (Markova 1990: 191).

It is precisely such creative interactions, constantly assimilating and 'reworking' reality, in other words, which produce, simultaneously, the dazzling diversity of cultural development *and* of individual development – diversity which far outstrips that which could be produced by randomly varying genes, or at any other regulatory level alone. *This* is what appears to have been the specific 'achievement' of human evolution – a 'method' of production of new means of coping with a changing world which, by general acknowledgement, accounts for the unprecedented success of humans as a species. That 'method', as I hope you have been able to see, is a developmental and historical one *par excellence*.

A systems summary

Traditional views of the role of genes in cognitive abilities is that of cognitive structures coded in the genes, randomly varying, but more or less attenuated by the environment in the course of their maturation. There has been an equivalent tendency to describe the 'environment' as a bland unstructured system of independent factors. Even those who use terms like 'gene–environment interaction' or 'epigenesis' tend to be sneaking in the same view. This has always produced confusion and conflict, as well as the perpetuation of the theoretical 'Tower of Babel', described in Chapter 1. In addition, it has fostered an ideology of fatalism and pessimism about human abilities.

Here I have tried to show that 'gene coding' of abilities, arising from selection of random mutations, would be a quite inadequate means for dealing with the changeable environments with which complex animals have had to deal. The transgenerational process of adaptation through natural selection has been incorporated in new *intra*-generational systems of regulation. In the face of environmental change, these regulations have increasingly involved development (and subsequently the lifelong development of socio-cognitive abilities) as a creative, rather than merely an 'expressive', process. Indeed, it is now clear that the range of potential expression of any given level of regulation, such as the genetic or the epigenetic, is severely limited unless nested within a higher level of regulation. For example, epigenetic expressions are themselves strictly limited to a few phenotypic classes or a range of values (so-called 'norms of reaction'), unless they are nested in, say, cognitive regulations, under the 'supervision' of which they become less constrained, with continuously changing expressions.

It is in this context that we need to be far more careful than in the past in our use of the term 'innate'. Its use may be permissable when theorists are making a distinction between relatively 'closed' and relatively 'open' developmental pathways (Mayr 1974), as in 'innate' territorial behaviours in some animals, or 'innate' reflexes, or the regulations that lead to the reliable construction of body parts. The problem is that this usage is easily taken to mean 'in the genes', and thus a return to the old mysticism.

Likewise, we need to be wary of attempts simply to 'meld' the biological and social, as in Plunkett and Sinha's (1992: 212) view of 'biological processes of growth and development as being continuous with psychological development, and the human mind as a product of a developmental unfolding of biological potential in a given sociocultural context'. The point is that it is the *discontinuites* which are most interesting and significant. The *intra*-generational regulations of epigenesis create a crucial discontinuity with the *inter*-generational process of genetic selection as mechanisms of adaptation. Similarly, cognitive and socio-cognitive regulations mark a discontinuity with epigenetic regulations. In any case we need to be aware that a term like 'innate cognitive ability' is a hopeless contradiction in terms, cognitive regulations having evolved to deal with circumstances for which 'innate' systems are quite inadequate.

A systems view also has important implications for understanding *variation* in cognitive regulations. What we 'see' in children's cognitive abilities, for example, may give the superficial appearances of variations in 'fixed' personal attributes like height, arm-length or physical strength. This is a common misconception. Indeed, psychologists have indulged the same misconception by constructing 'intelligence' and other tests of cognitive ability *as if* these characters are fixed attributes in that sense (as we shall see in the next chapter). The very essence of each succeeding level of regulation is the increasingly rapid production of adaptive variation for which its predecessor was inadequate. Cognitive abilities in any individual, therefore, may be better described as 'steady states' in a dynamic, self-organising system – often durable, no doubt, but still changeable, given certain conditions.

The systems view of 'productive' hierarchies contrasts with the view of hierarchies widespread in biology and psychology where each level is related to others, either by series of inclusiveness or of 'command lines' under an 'overall controller' (see e.g. Dawkins 1976). Technically, they fall in the category of *hyperstructure* in which the activities and products of any one level are conditioned by those in which it is nested by both feedback and feedforward processes (and so on, through several levels). Baas (1994) has offered a more formal description of the properties of hyperstructures compared with traditional hierarchies. As both Piaget and Baas have pointed out, such structures can be highly creative with the information they represent, generating new potentials and processes not prefigured in mere additive inclusions of levels: as Susan Oyama (1985) put it, there is development or *ontogeny* of information (see also Rose 1992). The general idea of more recently evolved levels of regulation conditioning and 'calling up' resources from those at more primitive levels is shown in Figure 3.8.

Steven Rose (e.g. 1981) has long advocated a 'levels', or hierarchical, systems view, as an antidote to reductionism or the collapsing of all the regulations that have evolved back into a shallow genetic determinism. An important consequence of such a view is how it alters our perception of 'causes' of developmental outcomes. Thus, instead of the activities of subordinate levels 'causing' outcomes at a superordinate level, they are best seen as furnishing the resources for the autonomous, self-organising functions *at* the higher level. Gene products are recruited as resources by epigenetic regulations; epigenetic processes are recruited as resources for lifelong plastic phenotypes, and so on. Naturally, any deficiency in such furnishing (such as a deleterious single-gene mutation) will result in a defect at a higher regulatory level. But this not the same thing as saying, in a far more general sense, that those regulations are 'genetically determined'. It is the failure to distinguish such arguments in the context of

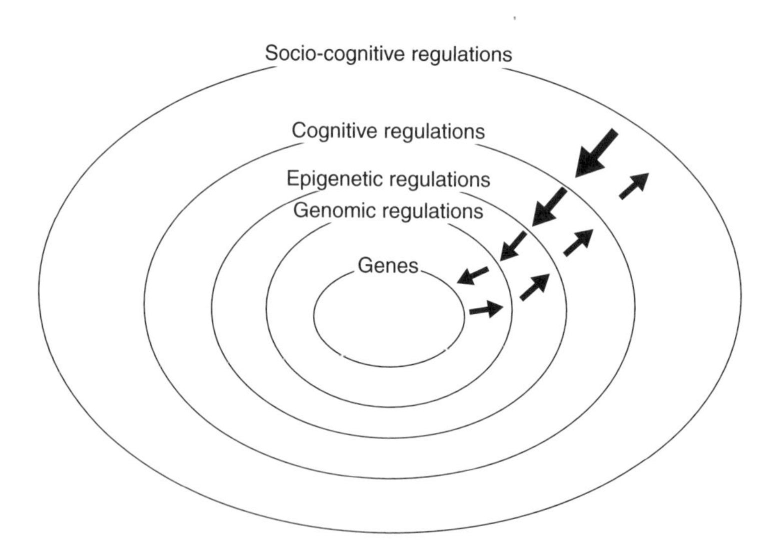

Figure 3.8 General picture of nested levels of regulation. Thickness of arrows stresses the point that regulation can be 'top-down' as well as 'bottom-up'. (Cognitive regulations are presented as the most evolved of several types of lifelong plastic regulations.)

multiple regulatory levels that has caused endless confusion in the nature–nurture debate.

There is, of course, abundant evidence for such 'single-gene' effects in humans (Wahlstrom 1990). Fortunately, they affect a very small proportion of people. The fact that many, if not most, single mutations have devastating effects on cognitive regulations – indeed, most are thought to result in such a loss of viability as to result in spontaneous abortion soon after conception – again points to the fact that these regulations depend on highly evolved, harmonious, gene systems. None the less, it becomes tempting to use such single-gene effects as a 'model' for the role of genes in the more general variation found in cognitive characters – i.e. to claim that the variation can be related to a simple sum of single-gene effects.

For example, Plomin *et al.* (1994: 108) argue that, 'rare alleles that drastically disrupt cognitive development are likely to be just the most easily noticed tip of the iceberg of genetic variability. It seems reasonable to expect that many more alleles nudge development up as well as down . . . and do not show such striking effects'. This seems to be suggesting that population variation in any character is made up of random sums of alleles, and that drastic disruptions (as in single-gene mutations) is just one tail of a continuum of such allelic variation. The alternative view is that there is no such 'iceberg': such effects are an isolated class of developmental consequences in evolved, cooperative gene systems.

I hope to have shown in this chapter that genes have a far more creative role in the development of systems of adaptable variation in complex, changing, environments than that of crude diversification of DNA. Of course, there may be special reasons why such 'raw' genetic variation may have persisted, even in cognitive abilities, in defiance of normal evolutionary forces. But these reasons have never been given. Unfortunately, investigators seem to want to have it both ways, and apply assumptions about unselected characters to, perhaps, the most intensively selected character there has ever been. In Chapter 5 I scrutinise these and many other (often mutually contradictory) assumptions in studies attempting to demonstrate genetic variation for cognitive ability in the guise of IQ. Before that, though, and because IQ has become so influential as a 'measure' of cognitive ability, I look at the assumptions underlying the use of that instrument.

References

Baas, N.A. (1994). Emergence, hierarchies and hyperstructures. In C.G. Langton (ed.) *Artificial Life III (Santa Fe Institute Studies in the Science of Complexity).* New York: Addison-Wesley.

Barkow, J.H., Cosmides, L. and Tooby, J. (1992). *The Adapted Mind.* Oxford: Oxford University Press.

Bates, E.A. and Elman, J.L. (1993). Connectionism and the study of change. In M.H. Johnson (ed.) *Brain Development and Cognition.* Oxford: Blackwell.

Bateson, P. (1988). The active role of behaviour in evolution. In M.-W. Ho and S.W. Fox (eds) *Evolutionary Processes and Metaphors.* Chichester: Wiley.

Blakemore, C. and Van Sluyters, R.C. (1975). Innate and environmental factors in the development of the kitten's visual cortex. *Journal of Physiology*, 248, 663–716.

Bozdogan, H. (1990). On the information-based measure of covariance complexity and its application to the evaluation of multivariate linear models. *Communications in Statistics: Theory and Methodology*, 19, 221–278.

Brockmeier, J. (1996). Construction and interpretation: exploring a joint perspective on Piaget and Vygotsky. In A. Tryphon and J. Voneche (eds) *Piaget–Vgotsky: The Social Genesis of Thought.* Hove: Psychology Press.

Bronfenbrenner, U. (1989). Ecological systems theories. *Annals of Child Development*, 6, 187–249.

Bruner, J.S. (1983). *Child's Talk.* Oxford, Oxford University Press.

Butterworth, G. and Bryant, P. (eds) (1989). *Causes of Development.* Brighton: Harvester Wheatsheaf.

Byrne, R. (1995). *The Thinking Ape: Evolutionary Origins of Intelligence.* Oxford: Oxford University Press.

Cole, M. (1988). Cross-cultural research in the sociohistorical tradition. *Human Development*, 31, 137–152.

Colwell, R.K. (1974). Predictability, constancy and contingency of periodic phenomena. *Ecology*, 55, 1148–1153.

Darwin, C. (1859). *On the Origin of Species by Means of Natural Selection.* London: Dent.

Dawkins, R. (1976). Hierarchical organisation: a candidate principle for ethology. In P.P.G. Bateson and R.A. Hinde (eds) *Growing Points in Ethology.* Cambridge: Cambridge University Press.

Dawkins, R. (1989). *The Selfish Gene.* 2nd edn. Oxford: Oxford University Press.

Dodson, S. (1989). Predator-induced reaction norms. *BioScience*, 39, 447–452.

Doise, W. (1988). On the social development of the intellect. In K. Richardson and S. Sheldon (eds) *Cognitive Development to Adolescence.* Hove: Erlbaum.

Donald, M. (1991). *Origins of the Modern Mind.* Cambridge, MA: Harvard University Press.

Dunn, J. (1988). *The Beginnings of Social Understanding.* Oxford: Blackwell.

Geertz, C. (1962). The growth of culture and the evolution of mind. In I.M. Scher (ed.) *Theories of Mind.* New York: Free Press.

Glaser, R, (1984). Education and thinking: the role of knowledge. *American Psychologist*, 39, 1–26.

Goodwin, B. (1985). Constructional biology. In G. Butterworth, J. Rutkowska and M. Scaife (eds) *Evolution and Developmental Theory.* Brighton: Harvester Wheatsheaf.

Goodwin, B. (1988). Morphogenesis and heredity. In M.-W. Ho and S.W. Fox (eds) *Evolutionary Processes and Metaphors.* Chichester: Wiley.

Goodwin, B. (1991). *Development.* Sevenoaks: Hodder and Stoughton.

Gottlieb, G. (1991). Experiential development of behavioural development: theory. *Developmental Psychology*, 27, 4–13.

Greenough, W.T., Black, J.E., and Wallace, C.S. (1993). Brain adaptation to experience. In M.H. Johnson (ed.) *Brain Development and Cognition.* Oxford: Blackwell.

Humphrey, N.K. (1976). The social function of intellect. In P.P.G. Bateson and R.A. Hinde (eds) *Growth Points in Ethology.* Cambridge: Cambridge University Press.

Huttenlocher, P.R. (1993). Morphometric study of human cerebral cortex development. In M.H. Johnson (ed.) *Brain Development and Cognition.* Oxford: Blackwell.

Jerison, H. (1973). *Evolution of the Brain and Intelligence.* London: Academic Press.

Johnson-Laird, P.N. (1983). *Mental Models.* Cambridge: Cambridge University Press.

Johnston, T.D. (1982). Learning and the evolution of developmental systems. In H.C. Plotkin (ed.) *Learning, Development and Culture.* New York: Wiley.

Johnston, T.D. and Gottlieb, G. (1990). Neophenogenesis: a developmental theory of phenotypic evolution. *Journal of Theoretical Biology*, 147, 471—495.

Kollar, E.J. and Fisher, C. (1980). Tooth induction in chick epithelium: expression of quiescent genes for enamel synthesis. *Science*, 207, 993–995.

Lancaster, J. B. (1975). *Primate Behaviour and the Emergence of Human Culture.* New York: Holt, Rinehart and Winston.

Lively, C.M. (1986). Predator induced shell dimorphology in the acorn barnacle *Chthamalus anisopoma. Evolution*, 40, 232–242.

Lovejoy, C.O. (1981). The origins of man. *Science*, 211, 341–352.

McGaugh, J.L., Bermudez-Rattoni, F., and Prado-Alcala, R.A. (1995). *Plasticity in the Central Nervous System.* Hove: Erlbaum.

Mackay, D.M. (1986). Vision – the capture of optical covariation. In J.D. Pettigrew, K.J. Sanderson, and W.R. Levick (eds) *Visual Neuroscience.* Cambridge: Cambridge University Press.

Markova, I. (1990). Causes and reasons in social development. In G. Butterworth and P. Bryant (eds) *Causes of Development.* Brighton: Harvester Wheatsheaf.

Marti, E. (1996). Mechanisms of internalisation and externalisation of knowledge. In A. Tryphon and A. Vonèche (eds) *Piaget–Vygotsky: The Social Genesis of Thought.* Hove: Psychology Press.

Mayr, E. (1970). *Population, Species and Evolution.* Cambridge, MA: Belknap Press.

Mayr, E. (1974). Behavior programs and evolutionary strategies. *American Scientist*, 62, 650–659.

Meyer, A. (1987). Phenotypic plasticity and heterochrony in *Cichlasoma managuense* (Pisces: Cichlidae) and their implications for speciation in cichlid fishes. *Evolution*, 41, 1357–1369.

Moran, N.A. (1992). The evolutionary maintenance of alternative phenotypes. *American Naturalist*, 139, 971–989.

Nederbragt, H. (1997). Hierarchical organisation of biological systems and the structure of adaptation in evolution and tumorigenesis. *Journal of Theoretical Biology*, 184, 149–156.

Oyama, S. (1985). *The Ontogeny of Information.* Cambridge: Cambridge University Press.

Parker, S.T. and Gibson, K.R. (1979). A developmental model for the evolution of intelligence and language in hominids. *Behavioural and Brain Sciences*, 2, 367–408.

Piaget, J. (1980). *Adaptation and Intelligence.* Chicago: University of Chicago Press.

Piaget, J. (1988). Piaget's theory. In K. Richardson and S. Sheldon (eds) *Cognitive Development to Adolescence.* Hove: Erlbaum.

Pittendrigh, R. (1958). Adaptation, natural selection and behaviour. In A. Roe and G.G. Simpson (eds) *Behavior and Evolution.* New Haven, CT: Yale University Press.

Plomin, R., McClearn, G.E., Smith, D.L., Vignetti, S., Chorney, M.J., Chorney, K., Venditti, C.P., Kasarda, S., Thompson, L.A., Detterman, D.K., Daniels, J., Owen, M., and McGuffin, P. (1994). DNA markers associated with high versus low IQ: the IQ Quantitative Trait Loci (QTL) project. *Behavior Genetics*, 24, 107–118.

Plotkin, H.C (1988). The evolution of closed and open programmes of development. In D.R. Garrard and J.D. Feldman (eds) *Development in the Neocortex.* Cambridge: Cambridge University Press.

Plotkin, H.C. (1994). *The Nature of Knowledge.* London: Penguin.

Plotkin, H.C. and Odling-Smee, F.J. (1979). Learning, change and evolution: an inquiry into the teleonomy of learning. *Advances in the Study of Behavior*, 10, 1–42.

Plunkett, K. and Sinha, C. (1992). Connectionism and developmental theory. *British Journal of Developmental Psychology*, 10, 209–254.

Purves, D. (1994). *Neural Activity and the Growth of the Brain.* Cambridge: Cambridge University Press.

Richards, G. (1987). *Human Evolution.* London: Routledge and Kegan Paul.

Richardson, K. (1992). Covariation analysis of knowledge representation: some developmental studies. *Journal of Experimental Child Psychology*, 53, 129–150.

Richardson, K. and Webster, D.S. (1996). Recognition of objects from point-light stimuli: evidence of covariation structures in conceptual representation. *British Journal of Psychology*, 87, 1–26.

Rogoff, B. and Chavajay, P. (1995). What's become of research on the cultural basis of cognitive development? *American Psychologist*, 50, 859–877.

Rollo, D.C. (1995). *Phenotypes: Their Epigenetics, Ecology and Evolution.* London: Chapman and Hall.

Rollo, D.C. and Shibita, D.M. (1991). Resilience, robustness and plasticity in a terestrial slug, with particular reference to food quality. *Canadian Journal of Zoology*, 69, 978–987.

Rose, S. (1972). *The Conscious Brain.* London: Weidenfeld & Nicolson.

Rose, S. (1981). From causations to translations: what biochemistry can contribute to the study of behaviour. In P.P.G. Bateson and P.H. Klopfer (eds) *Perspectives in Ethology*, Vol. 4. New York: Plenum Press.

Rose, S. (1992). *The Making of Memory.* London: Bantam.

Scaife, M. (1985). The implications of a structuralist biology for developmental psychology (commentary on Goodwin). In G. Butterworth, J. Rutkowska, and M. Scaife (eds) *Evolution and Developmental Theory.* Brighton: Harvester Wheatsheaf.

Scharloo, W. (1989). Developmental and physiological aspects of reaction norms. *BioScience*, 39, 465–472.
Shapiro, D.Y. (1981). Serial female sex changes after simultaneous removal of males from social groups of a coral reef fish. *Science*, 209, 1136–1137.
Singer, P. (1991). The functions of the neocortex. In J.C. Eccles and O. Creutzfeldt (eds) *The Principles of Design and Operation of the Brain*. Berlin: Springer-Verlag.
Singer, W. (1990). Search for coherence: a basic principle of cortical self-organisation. *Concepts in Neuroscience*, 1, 1–26.
Stearns, S.C. (1989). The evolutionary significance of phenotypic plasticity. *BioScience*, 39, 436–447.
Sur, M. (1993). Cortical specification: microcircuits, perceptual identity, and an overall perspective. *Perspectives on Developmental Neurology*, 1, 109–113.
Van Embden, M.H. (1971). *An Analysis of Complexity*. Amsterdam: Mathematical Centre Tracts 35.
Von Neumann, J. (1966). *Theory of Self-reproducing Automata* (ed. A.W. Burks). Urbana: University of Illinois Press.
Vygotsky, L.S. (1962). *Thought and Language* (trans. E. Hanfman and G.Vakar). Cambridge, MA: MIT Press.
Vygotsky, L.S. (1978). *Mind in Society* (ed. M. Cole, V. John-Steiner, S. Scribner, and E. Souberman). Cambridge, MA: Harvard University Press.
Vygotsky, L.S. (1981). The genesis of higher mental functions. In J.V. Wertsch (ed.) *The Concept of Activity in Soviet Psychology*. Armonk, NY: Sharpe.
Wachs, T.D. (1992). *The Nature of Nurture*. London: Sage.
Waddington, C.D. (1957). *The Strategy of the Genes*. London: Allen & Unwin.
Waddington, C.D. (1960). *The Ethical Animal*. Chicago: University of Chicago Press.
Wahlstrom, J. (1990). Gene map of mental retardation. *Journal of Mental Deficiency Research*, 34, 11–27.
Weinberger, N.M. (1995). Dynamic regulation of receptive fields and maps in the adult sensory cortex. *Annual Review of Neuroscience*, 18, 129–158.

4

IQ AND THE ELUSIVE COGNITIVE 'POWER'

Introduction

Historically, every age appears to have needed some mechanism of appeal to mystical, supra-human, forces for settling social issues. In the Middle Ages, trial by battle was common. By having disputees fight to the death, Divine Judgement was seen to intercede over rights and priorities in situations where *human* judgement could always be challenged as fallible. Indeed, as Hanson (1993: 29) points out, 'No legal procedure was more connected with feudalism or its spirit more thoroughly than the wager of battle'.

A constant problem in modern societies has been how to make judgements about people's rightful place in hierarchical role divisions, a problem that is particularly strong in the 'allocating' institutions of employment and education. In meeting this problem the idea of IQ as an overall mental power, with its marked *inequality* determined by biological forces beyond human control, has had a crucial part to play. Indeed, because the idea and instrumentation of IQ concentrate all the assumptions about quasi-biological, supra-human, forces discussed in previous chapters, I suspect that little, if anything, can be said to reflect the spirit of the modern age and its institutions more succinctly. This is why it has become the most popular view of human cognitive potential, and why I want to discuss it in detail here.

It needs to be mentioned at the outset, though, that the idea of IQ occupies a somewhat eccentric position, scientifically, in spite of its widespread support among psychologists and the general public. There have been, and still are, psychologists who have examined knowledge, learning, reasoning, and so on, in the way that all natural phenomena have been investigated, by painstaking systematic observations and detailed theory construction, followed by experimental testing. Jean Piaget is perhaps the outstanding modern example of such a theorist. IQ belongs to a more 'practical' or pragmatic approach. By simply equating variation in knowledge and cognition with a single underlying 'power' or 'strength' (which became called 'intelligence'), and then 'objectively' measuring it, it makes implications for social decision-making immediately, and infallibly, available.

This simple pragmatic view is, of course, the Platonic one described in Chapter 1. Although the IQ test has existed for nearly all of this century, and in spite of the dominance just mentioned, most people will know that there is still considerable disagreement about the meaning of IQ and the utility of the test. In the aftermath of yet another heated debate (that following the publication of the Herrnstein and Murray's *The Bell Curve* in 1994) the American Psychological Association made the brave, if perhaps optimistic, effort to summarise 'knowns and unknowns' in the intelligence debate (Neisser *et al.* 1996). In keeping with an effort to examine assumptions underlying beliefs and statements about cognitive abilities, I shall have occasion several times in the course of this chapter to refer to their deliberations. It is worth noting immediately, though, how the APA group concluded that the 'only recommendation we shall make is for further research and calmer debate' (Neisser *et al.* 1996: 78) and that 'We should be open to the possibility that our understanding of intelligence in the future will be rather different from what it is today' (p. 80). I hope to show that such an understanding is very much needed.

The origins of IQ testing

The modern psychometric approach undoubtedly begins with Sir Francis Galton in the late nineteenth century. The Platonic assumptions employed by Galton were starkly revealed in his explicit claims that the marked social grades in Britain were the inevitable expressions of different biological endowments, and he repeatedly condemned 'pretensions of natural equality' (1869: 56). This belief led Galton, like Plato, to favour a eugenic breeding programme for the improvement of society, and he wanted scientific measurement of 'natural ability', or intelligence, to further that end. Such measures would serve, he argued, 'for the indications of superior strains or races, and in so favouring them that their progeny shall outnumber and gradually replace that of the old one' (1883: 71).

So variation in cognitive ability, or 'intelligence', was reduced to biological variation. We simply need to measure it to make the essentially biological interventions (eugenic measures) that social improvement requires. But how was Galton to measure this intelligence in real individuals? Because he viewed this 'natural ability' as an overall power by analogy with physical strength, he reasoned that it must be manifest in all individual activities, even quite simple sensorimotor tasks. Thus we could measure this intellectual power by assessing individuals' performance on quite simple sensory and motor tasks. Galton and his associates, such as J.

McKeen Cattell, devised a number of such tasks. Here is a set devised by Cattell in 1890 (cited by Miller 1962):

1 *Dynamotor pressure.* How tightly can the hand squeeze?
2 *Rate of movement.* How quickly can the hand move through a distance of 50 cms?
3 *Sensation-areas.* How far apart must two points be on the skin to be recognised as two rather than one?
4 *Pressure causing pain.* How much pressure on the forehead is necessary to cause pain?
5 *Least noticeable difference in weight.* How large must the difference be between two weights before it is reliably detected?
6 *Reaction time for sound.* How quickly can the hand be moved at the onset of an auditory signal?
7 *Time for naming colours.* How long does it take to name a strip of ten coloured papers?
8 *Bisection of a 10 cm line.* How accurately can one point to the centre of an ebony rule?
9 *Judgement of ten-second time.* How accurately can an interval of ten seconds be judged?
10 *Number of letters remembered on once hearing.* How many letters, ordered at random, can be repeated exactly after one presentation?

But here we encounter a problem that has dogged the psychometric approach throughout its history. Most measures devised by humans can be shown to correspond with something existing in real life, either because we can sense it directly (e.g. length or weight) or through some intermediate measure that has a clear theoretical connection with that 'something' (e.g. concentration of certain gases in a breathalyser test and the amount of alcohol in the blood stream). In the case of Galton's tests, how were we to know that people's performance on these sensory and motor tasks is actually measuring the 'something' – the 'indication of superior strains and races' – he claimed to be measuring, and not something else, such as some other form of cognitive variation, accidental learning, confidence, or motivation?

Galton's solution was to become the hallmark of the 'intelligence' test: the performance of individuals simply needs to be compared 'with an independent estimate of the man's powers' (quoted by Evans and Waites 1981: 37). Galton's 'independent estimate' was much the same as Plato's. A person's position on the social ladder could itself be a 'measure' of intelligence. If individuals' performances on the tests corresponds with such status then the tests are measuring intelligence. Or, as Hunt (1983: 141) put it, 'We are presumed to know who is intelligent and to accept a test as a measure of intelligence if it identifies such persons'.

This is, and remains, the fundamental assumption of the intelligence testing movement, and its patent circularity is what has dogged it throughout this century. You should note how this strategy short-cuts any theory about the actual entity to be measured and evades the need to describe any chain of causation. It is merely an assumption that is easily questioned. After all, most of the people Galton identified as of superior social rank in Victorian England had attained that rank from a privileged starting point, through *social* inheritance of wealth and status. Whereas, in the case of the breathalyser, we have a demonstrable theoretical connection between breathalyser gases (or even walking down a white line) and actual blood alcohol levels, this is not the case with intelligence, sensorimotor performance, and social status. But the 'correlative-assumption-without-causal-demonstration' has loomed large in all the controversy that has surrounded the IQ test since its inception.

In the event the tests did not 'work' as Galton and Cattell had hoped. Differences were found among individuals on such measures. But, when the measures were compared with social status, it was found that there was little or no relationship. In other words, because the measured differences did not parallel differences assumed to be the 'real' expression of intelligence, the tests were disappointing. In spite of this failure, however, the general strategy of employing a correlative assumption signals the origins of the intelligence testing movement. The development of the strategy into a successful test simply awaited the invention of different kinds of tasks and a different criterion to correlate them with.

Binet's test

The motives behind the measurement of intelligence in the twentieth century have been predominantly practical, social motives, not theoretical ones. There were many social pressures fuelling such motives in the early years of this century. Among these was the introduction, in many countries of Europe, of systems of compulsory education. These measures brought into schools enormous numbers of children who, for whatever reasons, did not appear to be responding as hoped. There was natural concern about this. In 1904, the French Minister for Public Instruction appointed a Commission to study how retarded children could best be taught. Among its recommendations was one that no child should be removed to a special school without a 'medico-pedagogical' examination to determine his or her ability to profit from teaching in an ordinary school. But how was this to be done?

A member of this commission was Alfred Binet who, with assistants Henri and Simon, had been studying mental development, and ways of assessing it, for over a decade. In the light of Galton's and Cattell's failures, Binet was determined to concentrate on 'higher' mental functions. But what could these be? And how could they be measured?

Binet simply thought of all conceivable mental attributes that could possibly be quantified: memory, imagination, attention, comprehension of sentences and synonyms, aesthetic judgements, moral judgements, speed of acquiring a motor skill, and so on (see Miller 1962 for discussion). When he came to the work of the Commission, Binet already had considerable experience in devising such tests.

As Miller (1962: 313) goes on to describe the subsequent work of Binet and Simon:

> They used a large battery of mental tests, some hard, some easy. Binet collected tests from everywhere, and his own mind bubbled over with ideas for others. To discover which tests were useful, he and Simon spent endless hours in the schools with the children, watching, asking, testing, recording. Each proposed test had to be given to a large number of children. If a test did not distinguish the brighter from the duller, or the older from the younger, it was abandoned. Tests that worked were retained, even though they often failed to conform with the theoretical principles Binet and Henri had announced ten years earlier. The memory tests worked. And the tests of comprehension worked – comprehension of words of statements, of concepts, of pictures. Binet did not retain the tests on the basis of a theory; he watched the children and let their behaviour decide which tests were good and which were irrelevant.

Note how, although Binet eschewed the 'biological strength' assumption, he nevertheless adopted the correlative assumption. Only the criterion correlate was different: actual achievement in school as assessed by teachers' impressions, rather than social rank. On this basis Binet and Simon produced their first 'Metrical Scale of Intelligence' in 1905. It contained thirty items, designed for children aged three to twelve years, arranged in order of difficulty. They were grouped according to the proportion of a large group of children of a given age that had passed them and according to how they distinguished between children whom teachers also assessed as 'bright' or 'dull'. Here are examples of some of the items:

- imitating gestures and following simple commands;
- naming objects in pictures;
- repeating spoken digits;
- defining common words;
- drawing designs from memory;

- telling how objects are alike ('similarities');
- comparing two lines of unequal length;
- putting three nouns or three verbs into a sentence;
- 'abstract' (comprehension) questions (e.g. 'When a person has offended you, and comes to offer his apologies, what should you do?);
- defining abstract words (by describing the difference between such words as 'boredom' and 'weariness', 'esteem' and 'friendship').

(For a fuller description of all the items see Wolf 1973: 179–183.)

The tester simply worked through the items with each child, until the latter could do no more. Performance was then compared with the average for the age group to which the child belonged. If a child could pass half the tests expected of a six-year-old, say, then the child was said to have a mental age of six. Binet used the difference between the mental age and the chronological age as an index of retardation. He considered two years to be a serious deficiency (Miller 1962).

Thus was born the first modern intelligence test. Within a few years, translations were appearing in many parts of the world (for reasons mentioned above). In 1912, Stern proposed the use of the ratio of mental age to chronological age to yield the now familiar intelligence quotient or IQ:

$$\text{IQ} = \frac{\text{mental age}}{\text{chronological age}} \times 100.$$

The intelligence which Binet tested

It is necessary to pause at this stage to reflect on the principles entailed in the construction of Binet's test. This is because misunderstandings about it have been the cause of so much conflict ever since, and these conflicts are, if anything, even stronger today.

Practical points

From a purely practical viewpoint, Binet's scale appeared to be a brilliant success. It was easily and quickly administered and it actually identified the children it was supposed to identify. It is doubtful whether it did this any better than, say, teachers could have done on the basis of their experience with the children (after all this was the only criterion, next to age discrimination, of acceptability of test items). But the test had an element of objectivity in that exactly the same scale was given to all children; they were therefore not being judged by any *particular* teacher, but, as it were, by teachers in general. And it afforded immediate comparability with children of the same or different ages; so that *degree* of retardation seemed to be indicated.

Theoretical points

What it was retardation *in* is quite a different point. Note that the test result provides no new *psychological* information; nothing that was not known about a child already. How could it? Binet largely disregarded the whole question of a model or theory about the phenomenon being assessed. Although his search for test items was systematic and painstaking, this alone does not make it scientific. As Miller (1962: 315) put it, 'he was not over-concerned with scientific purity; he had a practical problem he urgently wanted to solve, and he did whatever seemed necessary to solve it'. We have already mentioned how 'Binet did not retain the tests on the basis of a theory', and that, 'they often failed to conform with the theoretical principles'.

In fact there were not many theoretical principles available. In 1905 Binet and Simon wrote

> we must make known the meaning we give to this vague and very comprehensive word 'intelligence'. Almost all the phenomena that occupy psychology are phenomena of intelligence. . . . There is in intelligence, it seems to us, a fundamental agent the lack or alteration of which has the greatest import for practical life, and that is judgement, otherwise known as good sense, practical sense, initiative, the faculty of adapting oneself. To judge well, to understand well, to reason well, these are the essential springs of intelligence.
>
> (Binet and Simon 1905: 194)

A fundamental agent that enters into all practical life? Good sense, judgement, etc? These represented no theoretical advance, since intelligence had been more or less vaguely described as such for centuries. In any case, Binet was not concerned about whether or not his test items actually matched these theoretical qualities, as we have seen. His effort at measurement was not scientific research, aimed at identifying components, relations, etc., as in the usual course of theory construction and theory testing. His purpose was entirely the practical one of screening for educational retardation.

Because of the way in which items were selected for these extrinsic, rather than their intrinsic (psychological), qualities, other psychologists, such as Yerkes in America, were quick to point out that 'Even the most enthusiastic believer in the Binet scale and methods cannot hope to maintain the thesis that at each or even at two ages precisely the same forms or aspects of human behaviour are measured' (quoted by Fancher 1985a: 122–123). Yerkes pointed out that what we really should have is a scale on which all individuals were being measured for the same thing: 'The difference', he said, was that 'between a relatively unscientific procedure and one which is striving to fulfil the essential requirements of scientific method' (quoted by Fancher 1985a: 123).

Of course Binet *assumed* that there was a consistency in his procedures and that what was being discriminated in children of different mental ages was the 'intelligence' he had earlier characterised more or less vaguely. And many others have made the same assumption since. But as Howe (1988) points out, this is an error of logic. To find an IQ test a useful device, in the sense that tests scores correlate with teachers' estimates, or otherwise with school performance, is one thing. Suggesting that in a Binet test performance we have identified the underlying 'cause' is quite another. Howe (1988) likens this reasoning to declaring a factory is productive 'because it has high productivity'. But the circularity of the correlative assumption quickly became forgotten in the wake of the *social* uses for which the test was soon being deployed, especially in America.

The growth of IQ in America

Within a matter of years, as Miller (1962) points out, Binet's test was in use in many other parts of the world. The most rapid developments took place in the USA, where 'feeblemindedness', especially among the new wave of immigrants, was seen to be a pressing problem from the point of view of education and national social security. Previous attempts to assess feeblemindedness systematically had failed, as we saw earlier. All this changed when Henry H. Goddard translated Binet's test into English in 1910.

Goddard was Director of the Vineland Training School for Feeble-minded Boys and Girls in New Jersey. Like Galton (but unlike Binet), Goddard fully subscribed to the biological assumptions and believed that feeblemindedness reflected one tail of a continuum of genetically inherited factors. He argued that feebleminded people must not be allowed to reproduce and found the IQ test to be a crucial instrument in the propaganda he raised. Those who followed Goddard in the IQ testing movement advocated the sterilisation of the feebleminded, a policy which was actually adopted by many states in America, resulting in tens of thousands of surgical operations (Pickens 1970; Karier 1972; Kamin 1974). Binet was later to protest about the Anglo-American interpretation of the test as a measure of some fixed quantity of the individual (Hunt 1961).

In a ghastly tragi-comedy Goddard and associates targeted the waves of immigrants, pouring into the USA at that time, and managed to ensure that every one of them was given the IQ test as soon as they landed, using the tests in English through interpreters. In his account of the process, Goddard himself gives an ironic glimpse of the 'objectivity' of this testing when he points put how, 'After a person has had considerable experience in this work, he almost gets a sense of what a feebleminded person is so that he can tell one afar off' (cf. Locurto 1991: 17). By these means the country came to be told that 83 per cent of Jews, 80 per cent of Hungarians, 79 per cent of Italians, and 87 per cent of Russians were feebleminded. The amount of 'feeblemindedness' thus exposed soon had these psychologists pressing ardently for immigration controls, which eventually became law in 1924. As many historians of science have pointed out, the subsequent growth of IQ was that of a blatantly racist tool.

Writing in popular journals and magazines these psychologists thus came to present the IQ test as a test of the actual genetic worth of people. It could help rid the country of feeblemindedness, crime, industrial unrest, and social disorder, all of which were attributed to genes. As Terman, the author of the most famous IQ test devised, put it,

> It is safe to predict that in the near future intelligence tests will bring tens of thousands of these high-grade defectives under the surveillance and protection of society. This will ultimately result in curtailing the production of feeblemindedness and in the elimination of an enormous amount of crime, pauperism, and industrial inefficiency.
>
> (Quoted by Locurto 1991: 18)

By using his IQ test we could 'preserve our state for a class of people worthy to possess it' (Terman 1917: 162). It was a view which seeped into the American public psychology, and soon into that in Britain, where eugenics ideas had sprung and were being spread by psychologists.

The growth of IQ in Britain

Followers of Galton not only carried his gospel to America, but founded an influential campaign in the UK. The Eugenics Education Society was founded in 1907 and its journal, *Eugenics Review*, a year later (Lowe 1980). Its members were soon spreading the gospel about the degeneration of the 'race' in popular seminars, magazines, and the pages of newspapers including the columns of *The Times*. The general message was that 'sooner or later society must protect itself from the unrestricted propagation of tainted stocks' (Auden 1909, quoted by Lowe 1980: 3). While the psychologist Cyril Burt (1909) was arguing that intelligence was innate, and social class differences due to heredity, Karl Pearson was insisting, in the *Encyclopaedia Britannica*, that 'It is cruel to the individual, it serves no social purpose, to drag a man of only moderate intellectual power from the hand-working to the brain-working group' (quoted by MacKenzie 1979: 137). The new intelligence tests naturally suggested themselves as measures of such 'racial' and class stock, and in 1911 a report to the Board of Education recommended the use of them for the identification of defectives (Lowe 1980).

With the advice of psychologists like Cyril Burt the report of the Consultative Commission on Education of 1938 (the Spens Report) in Britain was able to declare:

> Intellectual development during childhood appears to progress as if it were governed by a single central factor, usually known as 'general intelligence' . . . . Our psychological witnesses assured us that it can be measured approximately by means of intelligence tests . . . it is possible at a very early age to predict with accuracy the ultimate level of a child's intellectual powers . . . . It is accordingly evident that different children . . . if justice is to be done to their varying capacities, require types of education varying in certain important respects
>
> (Consultative Commission on Education 1938: 357–358)

Thus was founded the British 11+ examination and the selective school system.

Terman's revisions of the Binet scales

In 1916 Lewis Terman at Stanford University published what he called the Stanford Revision of the Binet–Simon Intelligence Scale (hereafter Stanford–Binet). Like the Binet test it was based on a variety of items selected, not from any theoretical distinction about what tests what, but merely according to their ability to discriminate among children in a way which agreed with a *prior* discrimination – i.e. chronological age and teacher's judgements. But there were many more items (ninety in all) which, therefore, required more time to administer, and more elaborate instructions to test administrators. It very quickly became the 'standard' IQ test on both sides of the Atlantic. This test has been revised several times since, to include a still wider variety and greater number of items. Here are examples of items which a ten-year-old would be expected to pass.

- *Vocabulary.* Correctly defining words.
- *Block counting.* Counting the number of blocks in a three-dimensional picture in which some of the blocks are obscured.
- *Abstract words.* Defining words like 'pity' and 'curiosity'.

- *Finding reasons.* Explaining why, for example, children should not be late for school.
- *Word naming.* Naming as many words as possible in one minute.
- *Repeating digits.* Repeating in order a sequence of digits such as 4–7–3–8–5–9.

The Wechsler scales

David Wechsler produced an alternative test in 1939 (revised several times since) because he believed the Stanford–Binet to be unsatisfactory in form and content for assessing adults. In 1949 it was augmented by the Wechsler Intelligence Scale for Children (the WISC), revised in 1974; and by the Wechsler Pre-school and Primary Scale of Intelligence. For illustration, here is a brief description of the sub-scales of the WISC (used in the age range five to fifteen years):

A Verbal scale

1 *General information.* Answering questions like 'What is steam made of?'
2 *General comprehension.* Answering questions like 'What should you do if you see someone forget his book when he leaves his seat in a restaurant?'
3 *Arithmetic.* Solving typical school textbook problems in a given time; e.g. 'Three men divided eighteen golf balls among them. How many golf balls did each man receive?'
4 *Similarities.* Describing how two things are alike; e.g. 'lion' and 'tiger'.
5 *Vocabulary.* Defining words of increasing difficulty.
6 *Digit span.* Repeating strings of numbers.

B Performance scale

1 *Picture completion.* Identifying the missing parts in incomplete pictures.
2 *Picture arrangement.* Arranging pictures in proper order so that they tell a story.
3 *Block design.* Pictures of block structures to be reproduced by arranging coloured blocks.
4 *Object assembly.* Making a picture (e.g. a horse) out of scrambled jigsaw-like pieces.
5 *Mazes.* Tracing a way out of a maze.

In all such tests, a list of acceptable and unacceptable answers is provided for each item. The problem is that nobody knows what exactly is the 'intelligence' that enables some children to give the correct answer to such questions, but not other children (apart from the circular idea that it is a result of general intelligence). There is little doubt that the vast majority of children can answer other questions or do other tasks that require equivalent levels of reasoning (see further below). So what is special about these particular items? Nobody really knows, and all we have to rely on is the correlational assumption: Children who can answer them are also those who tend to be 'bright' in school according to teachers.

The Stanford–Binet and the WISC have been far and away the most popular tests on both sides of the Atlantic, and probably are to this day. For example, in a small survey in this country, Quicke (1982) found that the WISC and the Stanford–Binet were regularly used by 82 per cent and 42 per cent, respectively, of educational psychologists. Use of these tests has increased gradually over the years as already indicated. But the fact that by the 1960s no individual test of 'general intelligence' had been developed in Britain had been long lamented. This situation was remedied in 1965 when a research project was set up under the auspices of the British Psychological Society to develop a test to replace the Stanford–Binet and the Wechsler scales. This led to the British Intelligence Scales (Elliot 1975).

Group tests

In spite of the apparent success of the Binet test the fact that it was administered by one tester to one testee at a time soon proved a major drawback. Those who would use such tests wanted to test large numbers of people at a time, as with children in school or recruits in the armed forces. The matter came to a head with the entry of the USA into the First World War and the persuasiveness of Robert Yerkes over the need to test all army recruits both for mental defectiveness and 'for the classification of men in order that they may be properly placed in the military service' (quoted by Fancher 1985a: 118). Two paper-and-pencil tests were devised, which could be given to people in large numbers at a time: the Army Alpha for literates and the Army Beta for illiterates. As Fancher (1985a) notes, by early 1918 these were being administered at the rate of 200,000 per month.

As with the earlier tests, construction consisted of devising batteries of items which had an intuitive correspondence with what the constructor saw as intelligence, and then selecting those which trials showed would make up a test with the required overall properties. Very soon such tests became extremely popular. In Britain group tests were available from the early 1920s. Today many such tests are employed for personnel selection in, for instance, the Civil Service, the armed forces, and in commerce and industry. From the 1940s they were extensively used in educational selection. There is scarcely a person, in fact, who has not been administered such a test at one time or another.

The types of items used in group tests include the familiar range of general information, arithmetic, classification, sequences, opposites, and so on, which are simply administered in questionnaire format with multiple-choice answers to be 'ticked'. A very common item is some type of analogy, having the general form A is to B as C is to ? (select D from a range of alternatives):

> Teacher is to pupil as Doctor is to . . .
> (hospital, nurse, patient, medicine, child)

'Matrix' items are also very common in group tests (Figure 4.1).

As Burt (1977: 32–33) notes about such items, 'the square pattern or "matrix" can be enlarged to include nine or even sixteen related items. . . . If lines, dots and simple geometrical figures are employed, there is hardly any limit to the material that can be systematically compiled for such purposes'. The convenience of administration of these tests obviates the direct supervision of the psychologist. Even the medical practitioner, interested in mental assessments, 'can get together a number of supplementary tests for himself; he will find them in many ways easier to invent, easier to apply, and often more reliable in their results, than the verbal' (p. 33). Perhaps the best known of all group tests, in fact, is the Raven's

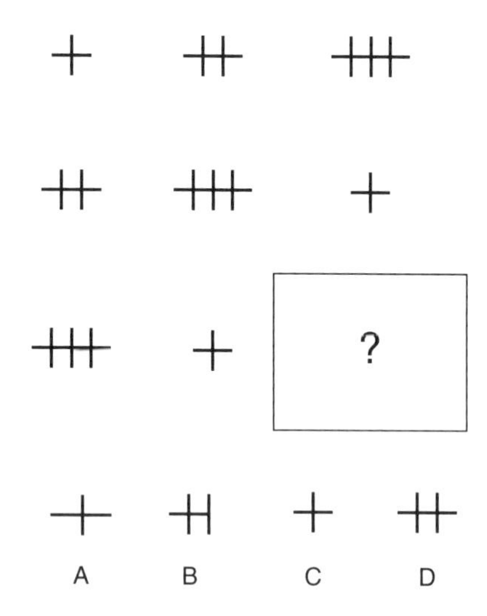

Figure 4.1 Example of a simple Matrix item. Testees have to select the missing element from the choices below.

Progressive Matrices, made up entirely of such items (see more about this below, pp. 124–126).

The proliferation of assumptions

One of the problems of becoming committed to strong assumptions is that, as they become manifest in real procedures and instruments, it becomes necessary to introduce others in order to sustain them. This proliferation of assumptions has happened a great deal with the construction and use of IQ tests, and I want to consider a few of them in this section. It is very important that readers understand this, because of the widespread impression that an IQ test is a 'direct' measure of a clear theoretical construct in the way that a breathalyser measures fitness to drive; as an instrument, it is something quite different from that. These new assumptions have all got to do with upholding the fundamental correlational assumption and the biological assumptions on which that, in turn, is based. They decide 'in advance', as it were, who is or is not intelligent, and then the superior scoring of such individuals becomes guaranteed by a long, drawn-out process known as 'item selection'.

For example, one of Terman's assumptions was that intelligence is an all-round, 'general' power, rather as Galton had envisaged. Consequently, each item was only selected if performance on it agreed to some extent with performance on the *whole sample* of items (i.e. if some children did well on a particular item but not on others already selected, then that item would not be included). This obviously produces the appearance of an 'agreement' between the items, that they are all measuring the same thing. This confirms, to some psychologists, that such a 'general' phenomenon is, indeed, being measured, although it is, of course, only an assumption built into the test.

The same applies to many other 'characteristics' of IQ. For example, the 'normal distribution, or bell-shaped curve, reflects (misleadingly as I have suggested in Chapters 1 to 3) key biological assumptions about the nature of cognitive abilities. It is also an assumption crucial to many statistical analyses done on test scores. But it is a property built into a test by the simple device of using relatively more items on which about half the testees pass, and relatively few items on which either many or only a few of them pass. Dangers arise, of course, when we try to pass this property off as something happening in nature instead of contrived by test constructors.

By a similar process we can assume that 'innate' differences between specific groups of people exist, and thus build such differences into the test as if the assumption is valid. Indeed any differences between groups of children can be built 'into' or built 'out of' the test, according to the test constructor's prior assumptions. We can illustrate this with the issue of sex differences. Boys tended to score a few points higher than girls in the original version of the Stanford–Binet, and at the time of the 1937 revision it was debated whether or not these should be allowed to persist. As McNemar, one of the researchers working on the revision explained,

> One who would construct a test for intellectual capacity has two possible methods of handling the problem of sex differences.
>
> 1 He may assume that all the sex differences yielded by his test items are about equally indicative of sex differences in native ability.
> 2 He may proceed on the hypothesis that large sex differences on items of the Binet type are likely to be factitious in the sense that they reflect sex differences in experience or training. To the extent that this assumption is valid, he will be justified in eliminating from his battery test items which yield large sex differences.
>
> The authors of the New Revision have chosen the second of these alternatives and sought to avoid using test items showing large differences in percents passing.
>
> (McNemar 1942: 56)

This is, of course, a clear admission of the subjectivity of such assumptions: while 'preferring' to see sex differences as undesirable artefacts of test composition, other differences between groups or individuals, such as different social classes or, at various times, different 'races', are seen as ones 'truly' existing in nature. Yet these, too, could be eliminated or exaggerated by exactly the same process of assumption and manipulation of test composition.

The process of self-fulfilment of assumptions is seen in yet another way. So far I have scarcely mentioned *development* of intelligence at all. In the psychometric approach this has never been much of an issue, and the reasons for that are fairly clear. Children's scores increase with age and this is seen simply as the maturation of the all-round strength or power that intelligence is conceived to be. But a dilemma is created by the kind of graphic representation of IQ shown in Figure 4.2 – i.e. 'growth curves' showing how individuals' scores increase with age. This suggests that the 'development' of IQ tails off after the age of fifteen years or so – in spite of the fact that most people would agree that our knowledge and our reasoning capabilities continue to increase. This is clearly a reflection of the assumption that children's intelligence increases steadily with age in keeping with school achievement, an assumption which is, again, built into the test by the process of item selection. The main problem is that we don't really know to what extent such 'growth curves' reflect something actually happening in children, or whether they are artefacts of test construction.

Figure 4.2 The development of intelligence according to the psychometric view – typical test scores (vertical axis) with age in years (horizontal axis).

This situation can be remedied by an equally subjective process of item selection. If we devise items that correlate with post-school achievement and blend these items in with the school-age items, we will produce an IQ that continues to 'develop' well beyond the school years.

Justification of assumptions

Psychologists have attempted to justify the use of these assumptions in a number of ways which may or may not bear close scrutiny. This is what I consider in this section.

One sense of justification is based on what is called the 'face validity' of test items. The general *psychological* content of the items has obviously engendered a good deal of respect about their genuineness. Yet this is not accompanied by any deeper theoretical rationale. Although the items obviously require *some* cognitive process or other, and although they can be given descriptive labels, such as 'block counting' or 'comprehension', these simply describe the visible content, not a cognitive process. As Terman himself (1942: 45) admitted, 'Certainly, tests which bear the same label are apt to be quite different as to content and as to the kind of ability called for'. As an analogy, I may, for example, feed you some juice, see that some internal system deals with it beneficially, and thus describe it as a 'juice digester'. But this is no advance whatsoever in our characterisation of digestion, nor in our understanding of other people who become sick when they drink the same liquid.

As another example of the false sense of security induced by 'face validity', consider performance on a 'spatial reasoning' item like block counting. A child who fails such an item may be perfectly adept at working out the whereabouts of her sock in the bottom left-hand drawer of the far chest, in the second room on the left, on the third floor of the house next door, and propelling herself in the appropriate directions to get it – yet in failing a simple spatial reasoning item be assessed as having little 'spatial reasoning'. So what is it that makes this everyday activity 'unintelligent' and performance on the items 'intelligent'? The only answer that psychometrists can give is that the item is passed by a higher percentage of children who also meet the other criteria like teachers' judgements of their intelligence – i.e. the correlational assumption once again.

A measure or test that is going to be used to make decisions about people, as in selection or screening, is most valid when the measure in question relates to a theory or causal model which identifies the component(s) being measured in the 'system' in question, and clearly describes the causal connection between those components and the value on the measure. Thus we can relate the colour of light on a breathalyser, through a chain of causation, to differences in a component in the blood stream which we can equally clearly relate to impaired judgement in driving. By being clearly related to the 'unseen' components, the test can be said to have high validity.

Psychometrists, in the absence of such theoretical description, simply reduce score differences, *blindly* to the hypothetical construct of 'natural ability'. The absence of descriptive precision about those constructs has always made validity estimation difficult. Consequently the crucial construct validity is rarely mentioned in test manuals. Instead, test designers have sought other kinds of evidence about the validity of their tests.

The validity of new tests is sometimes claimed when performances on them correlate with performances on other, previously accepted, and currently used, tests. This is usually called the *criterion* validity of tests. The Stanford–Binet and the WISC are often used as the 'standards' in this respect. Whereas it may be reassuring to know that the new test appears to be measuring the same thing as an old favourite, the assumption here is that (construct) validity has already been demonstrated in the criterion test.

In reality, the feelings of most psychologists about the validity of tests they use seems to rest on predictive correlations with school achievement and subsequent occupational level. As Jensen (1975: 346) put it, 'Intelligence tests have more than proved themselves as valid predictors of scholastic performance and occupational level'.

Correlations of IQ with school grades vary a great deal, but most fall within the range 0.4–0.6 (Block and Dworkin 1976; Brody 1985). A substantial part, at least, of such correlations is, of course, explained by the way in which items are selected *precisely* because performances on them correlate with performances in school, other items being rejected. As already mentioned several times, there is an obvious circularity in the argument. 'The fact that intelligence tests correlate with academic achievement and school progress is unquestioned. From the very way in which the tests were assembled it could hardly be otherwise' (Thorndike and Hagen 1969: 325).

There seem to be obvious logical problems when this wholly self-fulfilling correlation is taken to be the *best* evidence there is that IQ tests are giving information about something other than what we know about already. All that really happens, though, is that the formal appearance of the test and its numerical results create an illusion of tangibility to an idealistic concept. If predicting educational performance is the only reason for giving the test, why bother? The test cannot predict school performance any better than teachers' estimates against which they are calibrated. Even modern tests still take longer to administer than a short word from teachers which is at least as reliable a predictor. As Layzer points out,

> Admirers of IQ tests usually lay great stress on their predictive power. They marvel that a one-hour test administered to a child at the age of eight can predict with considerable accuracy whether he will finish college. But as Burt and his associates have clearly demonstrated, teachers' subjective assessments afford even more reliable predictors. This is almost a truism.
>
> (Layzer 1973: 238)

IQ scores correlate with school performance; but does this mean that they are measures of something tangible that *causes* school performance? Without clearer ideas of the cognitive nature of the mediation we are not entitled to claim this. Scientifically, this requires clear theory and experimental test. From a number of analyses, Ceci (1991) in fact argues that the direction of causation is precisely the opposite to that expected: i.e. experience of schooling or school-type culture makes children better at the school-type tasks that comprise IQ tests. But there is a host of candidates of such mediation, other than the mysterious general power. For example, if the knowledge and reasoning required in IQ test items reflect experience with a particular culture (as even a cursory survey of items would suggest), then a common cultural experience explains both school performance and IQ performance. This is no reflection on the 'powers' or 'potentials' of other children, other than that they have different cultural resources.

Similar claims and doubts surround the important issue of whether IQ scores predict competence outside of school, particularly future occupational level and job performance. Again, justification for IQ tests seems to involve a self-fulfilling ordinance. Since school performance determines the number of years we stay at school, which in turn determines the *level* at which we enter the job market, it is, likewise, unsurprising that a correlation is found between IQ test scores and occupational *level*.

But what about occupational *performance*? Do IQ test scores predict how well people do in their jobs? This is, after all, what we are really after in promoting children and young people through the system on the basis of test scores. If the correlational assumption is valid, and IQ scores measure a general cognitive ability independent of school achievement, then we would expect that 'cause' to persist in later life, including occupational performance.

The picture here becomes extremely cloudy. As Jensen (1970: 63) concludes, there are 'surprisingly low correlations between a wide variety of intelligence tests and actual proficiency on the job. Such correlations average about 0.20 to 0.25, and thus predict only four or five percent of the variance in work proficiency'. This has been confirmed in a number of more recent studies (Wagner 1994), some studies suggesting that, after correction for other confounding factors, the real correlation may be around zero (Ceci 1990). IQ does not reliably discriminate the more from the less competent in a wide range of domains examined (Ericsson and Charness 1994; Hulin, Henry, and Noon 1990). Likewise, with respect to the Raven's Matrices, often thought of as measuring the 'centre of gravity of intelligence' (see below, p. 126), Raven, Raven, and Court (1993: 41) have to conclude that 'the predictive validity of the RPM . . . to success within an occupation is . . . relatively low'.

As in so many other areas of IQ research, none of these correlations establishes any causal connection whatsoever. The APA group (Neisser *et al.* 1996: 83) note how such poor predictors are still taken as the best there is, although they also point out that no other factor has been seriously entertained or examined. That neglect may also reflect the power of the IQ/genetic-potential mystique. Remarkably, after reviewing the pessimistic nature of such correlations, the most recent manual for the Raven's Matrices concludes that

> the popular notion of General Ability – and with it the concept of the Mental Age that is embedded in the IQ and the educational and staff selection practices which are associated with 'Ability', 'Mental Age' and 'IQ' – does not merit the explanatory power and attention accorded to it by psychologists, managers, educators and educational therapists.
>
> (Raven, Raven, and Court 1993: 4)

Many authors have expressed their concern about the myths that the IQ testing enterprise may thus be sustaining. McClelland (1973: 2), for example, has warned us that 'the testing movement is in grave danger of perpetuating a mythological meritocracy in which none of the measures of merit bears significant demonstrable validity with respect to any measure outside of the charmed circle'. On similar grounds, Goodnow (1986: 88) has called for a closer examination of assumptions involved in interpreting such predictions: 'If we are to test for something, we should clearly have a clearer sense of what that something is. . . . Why not a moratorium until we know more clearly what we are trying to predict, and what the costs and benefits are of various ways of proceeding?' At the very least we surely need to beware of any claims that IQ reflects 'high cognitive ability' (e.g. Plomin and Thompson 1993) in any general or generally useful sense.

So what is being measured?

Psychology students are repeatedly told to scrutinise assumptions and never directly to interpret a correlation as a cause. Yet the whole IQ enterprise appears to be based on such assumptions and interpretations. So in IQ we are faced with a great theoretical void. The question 'What *is* intelligence?' is one which has plagued the psychometric approach. In this section I want to examine some possibilities that have been proposed.

This problem first came to light in a symposium in 1921 in which the editors of the *Journal of Educational Psychology* asked those prominent in the area of intelligence at that time to state what they considered 'intelligence' to be and by what means it could be best be measured by group tests. The diversity of answers received, and the absence of agreement among them, have been famous ever since. They led to the half-joking, half-exasperated claim that 'intelligence is what intelligence tests test' (Boring 1923: 36).

This exercise has recently been repeated by Detterman and Sternberg (1986). They wrote to a couple of dozen theorists, asking them the same questions that were put to the experts in 1921. Sternberg and Berg (1986) analysed the results for frequencies of mentioned attributes. Of the twenty-five attributes of intelligence mentioned in 1986, only three were mentioned by 25 per cent or more of respondents (half of the respondents mentioned 'higher level components'; 25 per cent mentioned 'executive processes'; and 29 per cent mentioned 'that which is valued by culture'). Over a third of the attributes were mentioned by less than 10 per cent of respondents in each case. The responses are in some ways similar to those given in the 1921 survey. Twenty-one per cent of respondents mentioned 'elementary processes (perception, sensation, attention)' as part of the definition of intelligence in both surveys. But there are some striking differences. For example, only 8 per cent of the 1986 respondents mentioned 'ability to learn', compared with 29 per cent in 1921.

In any domain of scientifically responsible and socially significant 'measurement' such results would be worrying, to say the least. If you asked mycologists to describe the attributes of edible mushrooms, and you received such a diverse set of responses, you might find it reasonable to conclude that they do not really know what mushrooms are, and that consumption of them is best avoided altogether. This might be worrying enough in the context of food consumption; *but these tests are being regularly used on our children, often with powerful conclusions about, and consequences for, them.*

The 'structure' of IQ

One attempt to theorise about 'what is being measured' has consisted of examining patterns of correlations among IQ test or other ability scores across large numbers of people. In the area of general school achievement, for example, children who do relatively well on one subject may tend to do relatively well on others. After examining such correlations in 1904, Charles Spearman saw them as evidence of an underlying general factor at work: '*there really exists a something that we may provisionally term . . . "General Intelligence"*' (quoted by Fancher 1985a: 342 – Spearman's emphasis). Declaring that the correlational approach signalled a 'Copernican revolution' in psychology, he went on to argue, 'In these principles we must venture to hope that the long missing genuinely scientific foundation for psychology has at last been supplied' (1923: 355; cf. Gould 1981). Psychologists from Burt (1909) to Jensen (1970) have hailed it as perhaps the most important discovery in the whole field of intelligence testing.

The results, themselves, have not gone unchallenged, though. Fancher (1985b: 345), for example, found Spearman's data and analyses to be 'marked from beginning to end by arithmetical mistakes, erroneous calculations of correlations, ambiguous or incomplete description of method, and inconsistencies of reporting'. Moreover, replications of Spearman's studies have suggested rather lower intercorrelations among different intellectual performances (see Fancher 1985a and b for details).

But Spearman's idea of an unseen general power behind the correlations, the 'pure' core of intelligence which he called '*g*', has been widely embraced ever since. As Spearman himself explained, 'The [*g*] factor was taken, pending further information, to consist in something of the nature of an "energy" or "power" which serves in common the whole cortex (or possibly, even, the whole nervous system)' (quoted in Fancher 1985a: 95). Thus 'intelligence' became reinforced as a mysterious, intangible strength or power, allegedly responsible for so much of the fate of humans, yet one which 'does not exist in the physical sense' (Eysenck 1991: 18). This is an example of what is known as 'reification', a tendency common to mysticism and theology, but one which we need to guard against in any science.

Over the last sixty years or so, however, psychometrists have continued to use Spearman's approach, or an elaboration of it known as factor analysis, with IQ test scores themselves. The very idea that we can get something 'out' of a pattern of test scores that has not been put 'in', in the course of test construction (see above on the assumptions used in item selection) – or that we are dealing with an entirely 'natural' measure – would appear naive, had it not attracted so much support. I will not go into this remarkable story here (see Gould 1981 and Richardson 1991a). Suffice it to say that investigators have simply made certain assumptions about 'what to expect' in the patterns of scores, and adjusted their analytical equations accordingly: not surprisingly, that pattern emerges!

Readers may be inclined to think that I am exaggerating or distorting here. But this procedure has actually gone on, and a range of patterns of 'factors', from the very same sets of scores, has emerged, according to the theorists' predilections, as if they were describing something in nature rather than assumptions in the investigators' heads, or built into the tests. Yet these different patterns of factors have been put to generations of students and teacher-trainees as 'models' of intelligence.

I need only illustrate with two such 'models': some theorists have preferred to 'find' a convergence of scores 'as if' reflecting a general factor (as already mentioned), while others have preferred to 'find' smaller clusters of scores 'as if' the product of distinguishable, separate abilities. Psychologists have seriously claimed to be telling us something about people in such analyses, on the basis of objective evidence, rather than simply constructing different stories out of different sets of assumptions. Thus, adherents of different assumptions can write in the very same volume about essentially the same data:

> Psychometric studies have now pretty well resolved this dispute: there clearly is need for a general factor to account for the 'positive manifold' usually produced when IQ scores are intercorrelated.
>
> (Eysenck 1986: 3)

> There are good reasons for discounting the idea that there is a single, unitary capacity of general intelligence. Most of the evidence before us suggests that humans have several different intellectual capacities for which there is no functional unity – or if there is, it has yet to be defined.
>
> (Horn 1986: 35)

In one recent study, based on still different assumptions, Carroll (1993) argued that over seventy different abilities could be identified in present tests! Thus, although IQ theorists claim to be measuring a function, they cannot describe it in any clear sense and cannot agree whether it is one function, several functions, or perhaps even dozens of functions. With so much variety of argument, based merely on the assumptions adopted, it is scarcely surprising that the APA Group (Neisser *et al.* 1996: 78) have to conclude that 'there is no full agreement on what g means: it has been described as a mere statistical regularity . . . a kind of mental energy . . . a generalised abstract reasoning ability . . . or an index measure of neural processing speed'.

Speed of information processing

Another recent attempt to pin down IQ to some meaningful causal variable has been to correlate IQ scores with either reaction time or 'inspection time'. In this latter task, subjects are presented with two lines of slightly different length, close together on a television screen, for very brief but slowly increasing periods (measured in fractions of a second), and asked to judge whether they are of the same or different lengths. The duration of presentation required (or 'inspection time') before individuals are able to make a judgement is considered by some psychologists to be a measure of their 'processing speed' (Nettlebeck 1987), and this, in turn, to be an index of 'neural efficiency'.

Correlations between inspection time and IQ scores, usually around 0.30, have been found. However, as the APA group points out, we do not know the basis of these correlations. Advocates of the view suggest some 'basic processing mechanism' such as 'quickness of intake of information' as the basis of individual differences in both inspection time and IQ (for review see Deary and Stough 1996), but the association could be due to confidence, familiarity, motivation, or attentional factors, for example. More generally, I hope readers will appreciate the irony in the attempted reduction of a complex regulatory system, which cognitive psychologists are still striving to understand, to a simple 'speed' variable. An analogy may be the explanation of individual differences in overall physiological metabolism by the rate at which we intake food.

Knowledge, context and culture

This preoccupation with 'speed' and 'power' reflects a particular mechanical view of the mind – one in which 'information' is represented and computations performed, as with data keyed into a machine. Such a view neglects the very *origins* of the cognitive apparatus as a device for constructing optimal representations of, and solutions to, reality, on the basis of information that is not only fragmentary, but also lies at several depths. Accordingly, it eschews any significant role for the knowledge background of individuals and the context in which reasoning is demanded. As Ceci and Roazzi (1992: 74) note, content and context in cognitive processing tend to be dealt with merely as sources of noise to be controlled.

On the contrary, though, many psychologists in recent years have been stressing that content and context are not just adjuncts to, or 'factors' in, cognitive performance; rather, they are *of the essence* of cognition (Butterworth 1992; Ceci and Roazzi 1992). This means that it is futile to attempt to separate process from content and context, and any claims that we are doing so in, say, an IQ test are seriously misleading. Bruner (1990; 1991) thus suggests that real-life cognitive tasks have narrative-like structures in which higher-order meanings furnish contexts in which sub-plots are conditioned. We might thus identify limiting factors in performance not so much in terms of intellectual incapacity, but as the individual's ability to generate a 'story line', or construct the salient events from the text.

Having evolved a rather brilliant device for constructing predictability from the deeper structures of situations, it is natural that humans attend to aspects of context and use their knowledge resources in dealing with life's problems. Indeed, such attentions and resources can be defined as the very essence of human intelligence. As a result, people will tend to map specific variables and covariations, and thus certain expected outcomes, even on to the most isolated details, wherever this is possible. For instance, it has been shown that children and adults will readily construct social and other interpretations from only three or four light points (so-called point light stimuli) moving on a computer screen (Richardson and Webster 1996b). In the case of an IQ item, different interactions between individual knowledge resources, the item, and the context in which it is formatted and presented, will mean that the crude information will become organised in different ways by different individuals, with different inferences or deductions. There are many empirical illustrations of such interactions in the literature.

For example, Ceci and Roazzi (1992) report a study in which children had to predict movements on a computer screen of three geometric shapes, actually dictated by an underlying complex rule imposed by the investigators. Even after 750 trials subjects were still predicting at just above chance. But performance dramatically improved when the shapes were substituted with animals (bees, birds, and butterflies) and incorporated into a game in which they had to be caught in a net, although exactly the same underlying rules of movement remained. In this context children were able to bring their background knowledge to bear on the problem, induce the underlying rules, and *deduce* accurate predictions from them.

Somewhat similarly, Goswami (1995: 890) has shown how performance on analogical reasoning tasks 'depends on the development of *relational knowledge* rather than the development of relational mapping ability per se' (Goswami's emphasis). Keil (1988) showed how performance on reasoning tasks can be dramatically improved if relations in background knowledge can be metaphorically mapped on to those in the task. Carey (1988) suggests the role of background knowledge in differential performance on class-inclusion tasks, and reviews studies indicating how children's scientific reasoning depends upon their degree of knowledge of the domain in question – a dependence that would, of course, be expected among scientists in general. It has also been shown that when reasoning problems identical in structure to those in the Raven's Matrices are couched in the framework of more general social schemas, then all children find them remarkably easy (Richardson 1991b; Richardson and Webster 1996a). Even in simple categorisation tasks, many studies have now shown how differences in background knowledge produce large differences in classification among the same elements (see Wattenmaker 1996, for review).

Constructors of tests like Raven's Matrices obviously assume that all of life's problems unfold before us as bland logical syllogisms, and we only need to apply (more or less rapid or powerful) reasoning processes to the premises to solve the puzzle. But real-life problems are far more complex in that predictability usually needs to be *constructed* from fragmentary information about associations that are usually conditioned across several hierarchical levels. Some of these levels will include human needs and goals (Donaldson 1978), which are themselves culturally conditioned.

The remarkable thing is that there is abundant anecdotal and empirical evidence that almost all children, from a very young age, are capable of far more complex reasoning than that demanded in even the more Advanced matrices. For example, children as young as three to four years can make inferences about novel goals given novel means, or vice versa, which are far more complex than the 'addition of figure' rule in a Raven's test (where, for example, a child sees one figure in position 1, two figures in position 2, and must infer three in position 3) (see Frye 1994 for review).

The 'rules' to be discovered in items like Raven's Matrices are not particularly complicated, in fact, when compared with everyday cognition. Studies of cognition in everyday situations such as workplaces discover far more complex cognitions (e.g. Scribner 1997). In one study, Ceci and Liker (1986: 255) studied betting among regular race-goers in the USA and found that it was 'a cognitively sophisticated enterprise, with experts using a mental model that contained multiple interaction effects and non-linearity'. Their performance, moreover, was totally unrelated to measured IQ.

Similar conclusions have been drawn from a long tradition of cross-cultural research on specific cognitive skills such as classification, logic, and memory: just as we would not consider administering 'verbal' items in a language different from that of the subjects themselves, so it seems naive to administer *non*-verbal items, the content of which ignores the knowledge concepts and cognitive structures in the culture in which the subjects have developed. It seems obvious that children growing up in a home and culture whose chief medium of economic survival and communication is printed text, standard oracy, and accountancy tools will have acquired the same tools to a greater extent than other children and will be better prepared for schooling *and* tests like the Raven's (see Rogoff and Chavajay 1995 for further discussion).

Testing for cultural affiliation

Science generally is involved in the characterisation of stable entities, or of the stable 'laws of motion' of those entities. A major problem is faced by any such attempt with human cognition, as in attempts to measure some inherent strength or power in it. This is that the cognitive system is itself a reactive creative system which can only, therefore, be described in relation to some context of activity (Brown, Metz, and Campione 1996). According to a socio-historical framework, knowledge concepts and cultural tools vary, not because of some idealistic cognitive 'power', but because of the different problem structures in the different classes and cultures in which people actually live. As Michael Cole explained (quoted by Luria 1976: xiv–xv), 'the structure of thought depends upon the structure of the dominant types of activity in different cultures' (see also Richardson 1991b). Accordingly, an alternative view of what is being 'measured' in IQ tests is that of degree of cultural affiliation – i.e. the extent to which children possess the knowledge and cognitions manifested in *one particular cognition-culture complex*, when there is a vast diversity of such complexes. This means, in effect, that they are testing for the background knowledge and cultural tools more prevalent in the culture of the test constructors.

In the most popular current individual tests, this cultural overload is pretty obvious. The vast majority (over 80 per cent) of the items in the Stanford–Binet and the WISC are verbal questions about background knowledge administered in a verbal examination setting with a superior person. There are non-verbal items, but these comprise a small part of both tests, not least because they have always been difficult to find. As Terman and Merrill (1961: 22) explained, regarding the 1960 Stanford–Binet, 'Many of the so-called performance test items tried out for inclusion in the scale were eliminated because they contributed little or nothing to the total score. They were not valid items for this scale'. What this means is that children of different cultures tended to perform equally well on them, so they had to be rejected (again, on statistical, not theoretical, grounds).

This point is illustrated most clearly in what is commonly seen as the quintessential IQ test, namely Raven's Matrices. Much weight has been placed on such items in debates because, being non-verbal, they give the superficial impression of being 'culture free', and are thus thought to measure the 'pure' core of general reasoning ability, close to the 'centre of gravity' of intelligence. The Raven's is thought to test 'the ability to reason and solve problems involving new information, without relying extensively on an explicit base of declarative knowledge derived from either schooling or previous experience' (Carpenter, Just, and Shell 1990: 404).

Yet it is not difficult to show that Raven's Matrices are at least as culturally tied as their verbal counterparts. First of all, the items are administered in a socially fraught 'test' situation, itself a specific cultural device. But so is the very layout of items. They are presented as black and white figures, flat out on paper, with the 'pick-up' of information arranged from top-left to bottom-right. These are culture-specific tools for handling information, more prominent in some groups than others.

But there is much more cultural specificity in the Raven's test than the general framework of the items. The very inductions required – 'constant in a row'; 'quantitative pairwise progression'; 'figure addition or subtraction'; and so on (Carpenter, Just, and Shell 1990) – are emphatically culture loaded, in the sense that they reflect further information-handling 'tools' for storing and extracting information from text, from tables of figures, or accounts, timetables, and so on, all of which are more prominent in some cultures and subcultures than others.

In other words, items like the Raven's contain a hidden structure which makes them *more*, not less, culturally steeped than any other kind of 'intelligence' testing item: 'the most *systematically acculturated* tests', as Keating and Maclean (1987: 243; Keating and Maclean's emphasis) put it. This cultural specificity is hardly surprising because, as already mentioned, the items are the products of the cognitions of human beings, themselves immersed in a specific cultural milieu. The idea that the Raven's somehow separates children on the basis of a general 'power', intrinsic to the child, rather than simply identifying their cultural background, seems an over-optimistic one. Furthermore, the kinds of 'cultural tools' just mentioned are likely to be those further acquired in schools. Little wonder that Ceci (1990) can argue that IQ status is largely a *result* of schooling not a cause of it: i.e. the direction of prediction is the *opposite* of that usually inferred.

When we add to these considerations the well-demonstrated differences in beliefs in cognitive self-efficacy across classes and cultures (for review see Bandura *et al.* 1996, and Chapter 6) the idea of the Raven's as a pure 'power' measure seems even more naive. It is perhaps hardly surprising that the Raven's (along with other IQ measures) predicts little by way of cognitive performance in the real world (Raven, Raven, and Court 1993).

The most reasonable answer to the question 'What is being measured?', then, is 'degree of cultural affiliation': to the culture of test constructors, school teachers and school curricula. It is (unconsciously) to conceal this that all the manipulations of item selection, evasions about test validities, and searches for *post hoc* theoretical underpinning seem to be about. What is being measured is certainly not genetically constrained complexity of general reasoning ability as such. The fact that the average IQ score in many countries has risen sharply over the last two generations might well testify to that (Flynn 1987); it accompanies significant demographic swelling in numbers of the middle class over that period. Consistent with the same interpretation is the observation that, when members of dominant and subordinate groups migrate to another country (as with Japanese to America), previous IQ 'gaps' between them completely disappear (Ogbu 1986; cf. Block 1995).

This is certainly not a new answer. It has been put many times in the course of the history of the IQ test. But it seems more valid than ever in the light of recent research on cognition in context (Ceci and Liker 1986; Light and Butterworth 1992; Richardson and Webster 1996a). This possibility is at least worthy of far more consideration, as any *detached* observation of children's remarkable cognitive abilities in their normal social or cultural context should have impressed upon us. Of course, when encountering children who speak our language, have similar knowledge and thought processes as ours, similar forms of self-presentation, and so on, we have a natural inclination to perceive and label them as 'brighter'. But to raise such subjectivism to the status of a universal ladder of human intelligence is misleading as well as unscientific. Yet this is precisely what has been achieved by the idea and instrumentation of IQ.

Summary of assumptions

IQ testing has its origins in many of the quasi-biological assumptions about cognitive ability described in Chapter 1. Chief among these is:

1 the assumption that cognitive ability varies as a general, underlying 'strength' or 'power' which is determined biologically by randomly varying genes.

Because the nature of this power has not been successfully theorised, attempts to measure it have relied on doing so indirectly through using the 'correlational assumption':

2 that some existing discrimination among individuals (e.g. social class, school achievement) can be taken as a 'measure' of intelligence, and that we can have a valid test of intelligence if it is constructed such as to 'agree' with such discrimination.

Both of these assumptions then give rise to a range of additional assumptions which become 'built in' to the tests:

3 that the intelligence 'power' is an all-round, general one, so that a person will tend to do equally well on all parts of a test and a single score tells all that we need to know about that person's cognitive ability;
4 that intelligence will vary like any other 'biological' character such that its distribution in a population will approximate a 'normal curve';
5 that some group differences (e.g. social class, 'race') in intelligence are 'natural' – and therefore to be reflected in IQ score differences – while others are not;
6 that, in keeping with the biological assumptions, intelligence will 'mature' in a linear fashion up to about eighteen years, like physical characters such as height, then level off;
7 that we can theorise about different 'structures' of intelligence simply by analysing correlations among IQ scores;
8 that, because of (1) above, it is possible to devise items that will tap this power irrespective of the knowledge background and other cultural contexts of subjects.

I scrutinised all the quasi-biological assumptions in Chapters 1 to 3 and found them to be wanting, even paradoxical (for example, many are assumptions relevant to a character that has not evolved). The other assumptions have been examined above and again found to be wanting.

Evaluation of these assumptions ought to give us pause for thought about any claims that IQ is a 'scientific' measure of general cognitive ability. These claims appear to be hollow, and even their strongest advocates have to admit to this occasionally. As Eysenck (1986: 4) explains, the emphasis on Binet-type scores led 'to a neglect of theoretical and experimental investigations of intelligence and hence to the (largely justified) accusation that intelligence testing is a technology not based on proper scientific foundations'. And yet psychologists everywhere make the practical assumption that we can use IQ tests, and make strong decisions about people on the basis of test scores, without further theoretical justification.

We surely need to remember, not only this lack of theoretical foundation: since IQ scores display only moderate correlations with educational achievement (in spite of efforts to build that in, and with less predictive power than teachers' judgements) and hardly any with subsequent occupational performance, IQ appears to have little practical use at all. Little wonder that some psychologists have called for a moratorium on their use.

So why do we cling to the IQ test so avidly? One possible answer may lie with the way that its assumptions and products fire the public imagination about the nature of ability (witness the popularity of popular texts with titles like *Test Your Own IQ) and* reinforce public policy regarding allocation to social roles, especially in education and employment. In other words, the IQ test has persisted because it has played an important *ideological* role in society. This was certainly true in the earlier part of this century when Binet's sensitive approach to devising a limited tool for educational screening became used in upholding privilege against a 'feebleminded' underclass or inferior 'races'. In modern times there seems little doubt that the same assumptions and products engender a belief in the cognitive mediocrity of most children and adults. As Scarr (1989: 76) puts it, 'A theory of general intelligence fits better a culture and period of relative pessimism about human perfectibility'.

References

Bandura, A. Barbaranelli, C., Caprara, G.V., and Pastorelli, C. (1996). Multifaceted impact of self-efficacy beliefs on academic functioning. *Child Development*, 67, 1206–1222.

Binet, A. and Simon, T. (1905). Méthodes nouvelles pour le diagnostic du niveau intellectuel des anormaux. *L'Année psychologique*, 11, 191–244.

Block, N. (1995). How heritability misleads about race. *Cognition*, 56, 99–128.

Block, N. and Dworkin, G. (eds) (1976). *The IQ Controversy.* New York: Pantheon.

Boring, E.G. (1923). Intelligence as the tests test it. *New Republic*, June, 35–37.

Brody, N. (1985). The validity of tests of intelligence. In B.B. Wolman (ed.) *Handbook of Intelligence*. New York: Wiley.

Brown, A.L., Metz, K.E., and Campione, J.C. (1996). Social interaction and individual understanding in a community of learners: the influence of Piaget and Vygotsky. In A. Tryphon and J. Vonèche (eds) *Piaget–Vygotsky: The Social Genesis of Thought.* Hove: Psychology Press.

Bruner, J.S. (1974). *Beyond the Information Given: Studies in the Psychology of Knowing.* London: Allen & Unwin.

Bruner, J.S. (1990). *Acts of Meaning.* Cambridge, MA: Harvard University Press.

Bruner, J.S. (1991). The narrative construction of reality. *Critical Inquiry*, Autumn, 1–21.

Burt, C. (1909). Experimental tests of general intelligence. *British Journal of Psychology*, 3, 94–177.

Burt, C. (1949). The structure of the mind: a review of the results of factor analyses. *British Journal of Educational Psychology*, 19, 176–199.

Burt, C. (1977). *The Subnormal Mind.* Oxford: Oxford University Press.

Butterworth, G. (1992). Context and cognition in models of cognitive growth. In

P.H. Light and G. Butterworth (eds) *Context and Cognition: Ways of Learning and Knowing*. Hemel Hempstead: Harvester Wheatsheaf.

Carey, S. (1988). Are children fundamentally different kinds of thinkers and learners than adults? In K. Richardson and S. Sheldon (eds) *Cognitive Development to Adolescence*. Hove: Erlbaum.

Carpenter, P.A., Just, M.A., and Shell, P. (1990). What one intelligence test measures: a theoretical account of the processing in the Raven Progressive Matrices Test. *Psychological Review*, 97, 404–431.

Carroll, J.B. (1993). *Human Cognitive Abilities: A Survey of Factor Analytic Studies*. Cambridge: Cambridge University Press.

Cattell, J. McK. (1890). Mental tests and measurement. *Mind*, 15, 373–380.

Ceci, S.J. (1990). *On Intelligence: More or Less*. Englewood Cliffs, NJ: Prentice-Hall.

Ceci, S.J. (1991). How much does schooling influence general intelligence and its cognitive components? A reassessment of the evidence. *Developmental Psychology*, 27, 703–722.

Ceci, S.J. and Liker, J.K. (1986). A day at the races: a study of IQ, expertise, and cognitive complexity. *Journal of Experimental Psychology: General*, 115, 255–266.

Ceci, S.J. and Roazzi, A. (1992). The effects of context on cognition: postcards from Brazil. In R.J. Sternberg and R.K. Wagner (eds) *Mind in Context*. Cambridge: Cambridge University Press.

Consultative Commission on Education (1938). *The Education of the Adolescent* (Spens Report) London: HMSO.

Deary, I.J. and Stough, C. (1996). Intelligence and inspection time. *American Psychologist*, 51, 599–608.

Detterman, D.K. and Sternberg, R.J. (1986). *What is Intelligence? Contemporary Viewpoints on its Nature and Definition*. Norwood, NJ: Ablex.

Donaldson, M. (1978). *Children's Minds*. London: Fontana.

Elliot, C.D. (1975). The British Intelligence Scale takes shape. *Education*, 25, 460–465.

Elliot, C.D., Murray, D., and Pearson, L.S. (1978). *The British Ability Scales*. Windsor: NFER.

Embretson (née Whitely), S. (1985). Review of *The British Ability Scales*. In J.V. Mitchell Jr (ed.) *The Ninth Mental Measurements Yearbook*, Vol. 1. Lincoln, NB: University of Nebraska Press.

Ericsson, K.A. and Charness, N. (1994). Expert performance: its structure and acquisition. *American Psychologist*, 49, 725–747.

Evans, B. and Waites, B. (1981). *IQ and Mental Testing: An Unnatural Science and its Social History*. London: Macmillan.

Eysenck, H.J. (1986). The theory of intelligence and the psychophysiology of cognition. In R.J. Sternberg (ed.) *Advances in the Psychology of Human Intelligence*, Vol. 3. Hillsdale, NJ: Erlbaum.

Eysenck, H.J. (1991). Letter. *The Psychologist*, 4, 17–18.

Fancher, R.E. (1985a). *The Intelligence Men: Makers of the IQ Controversy*. New York: Norton.

Fancher, R.E. (1985b). Spearman's original computation of g: a model for Burt? *British Journal of Psychology*, 76, 341–352.

Flynn, J.R. (1987). Massive IQ gains in fourteen nations: what IQ tests really measure. *Psychological Bulletin*, 101, 171–191.

Frye, D. (1994). Causes and precursors of children's theories of mind. In D.F. Hay and A. Angold (eds) *Precursors and Causes in Development and Psychopathology*. London: Wiley.

Galton, F, (1869). *Heredity Genius: An Inquiry into its Laws and Consequences.* London: Macmillan.

Galton, F. (1883). *Inquiry into Human Faculty and its Development.* London: Macmillan.

Gardner, H. (1984). *Frames of Mind: The Theory of Multiple Intelligences.* London: Heinemann.

Gardner, H. (1991). *The Unschooled Mind.* London: Fontana.

Goodnow, J.J. (1984). On being judged 'intelligent'. In P.S. Fry (ed.) *Changing Conceptions of Intelligence and Intellectual Functioning.* Amsterdam: North-Holland.

Goodnow, J.J. (1986). A social view of intelligence. In D.K. Detterman and R.J. Sternberg (eds) *What is Intelligence? Contemporary Viewpoints on its Nature and Definition.* Norwood, NJ: Ablex.

Goswami, U. (1995). Transitive relational mappings in three- and four-year-olds. *Child Development*, 66, 877–892.

Gould, S.J. (1981). *The Mismeasure of Man.* New York: Norton.

Hanson, F.A. (1993). *Testing, Testing: Social Consequences of the Examined Life.* Berkeley, CA: University of California Press.

Herrnstein, R.J. and Murray, C. (1994). *The Bell Curve.* New York: Free Press.

Howe, M.J.A. (1988). The hazards of using correlational evidence as a means of identifying the causes of individual ability differences: a rejoinder to Sternberg and a reply to Miles. *British Journal of Psychology*, 79, 539–545.

Horn, J. (1986). Intellectual ability concepts. In R.J. Sternberg (ed.) *Advances in the Psychology of Human Intelligence*, Vol. 3. Hillsdale, NJ: Erlbaum.

Hulin, C.L, Henry, R.A., and Noon, S.L. (1990). Adding a dimension: time as a factor in the generalizability of predictive relationships. *Psychological Bulletin*, 107, 328–340.

Hunt, E. (1983). On the nature of intelligence. *Science*, 219, 141–146.

Hunt, J. McV. (1961). *Intelligence and Experience.* New York: Ronald Press.

Jensen, A.R. (1970). Another look at culture-fair testing. In J. Hellmuth (ed.) *The Disadvantaged Child.* New York: Brunner-Mazel.

Jensen A.R. (1975). Race intelligence and genetics: the differences are real. In J.M. Whitehead (ed.) *Personality and Learning.* London and Milton Keynes: Hodder and Stoughton in association with The Open University Press.

Joynson, R.B. (1989). *The Burt Affair.* London: Routledge.

Kamin, L. (1974). *The Science and Politics of IQ.* New York and London: Wiley.

Karier, C. (1972). Testing for order and control in the corporate liberal State. *Educational Theory*, 22, 154–180.

Keil. F. (1988). On the structure-dependent nature of stages of cognitive development. In K. Richardson and S. Sheldon (eds) *Cognitive Development to Adolescence.* Hove: Erlbaum.

Keating, D.P. and MacLean, D.J. (1987). Cognitive processing, cognitive ability, and development: a reconsideration. In P.A. Vernon (ed.) *Speed of Information Processing and Intelligence*. New York: Ablex.

Layzer, D. (1973). Science or superstition? A physical scientist looks at the IQ controversy. *Cognition*, 1, 265–300.

Lewis, C. (1994). Episodes, events and narratives in the child's understanding of mind. In C. Lewis and P. Mitchell (eds) *Children's Early Understanding of Mind*. Hove: Erlbaum.

Light, P. and Butterworth, G. (eds) (1992). *Context and Cognition: Ways of Learning and Knowing*. Hemel Hempstead: Harvester Wheatsheaf.

Locurto, C. (1991). *Sense and Nonsense about IQ*. New York: Praeger.

Lowe, R. (1980). Eugenics and education: a note on the origins of the intelligence testing movement in England. *Educational Studies*, 6, 1–8.

Luria, A.R. (1976). *Cognitive Development: Its Cultural and Social Foundations*. Cambridge, MA: Harvard University Press.

McClelland, D.C. (1973) Testing for competence rather than for 'intelligence'. *American Psychologist*, 28, 1–14.

MacKenzie, D. (1979). Karl Pearson and the professional middle class. *Annals of Science*, 36, 125–143.

McNemar, Q. (1942). *The Revision of the Stanford–Binet Scales*. Boston: Houghton-Mifflin.

Miller, G.A. (1962). *Psychology: The Science of Mental Life*. Harmondsworth: Penguin.

Neisser, U., Boodoo, G., Bouchard, T.J. Jr, Boykin, A.W., Brody, N., Ceci, S.J., Halpern, D.F., Loehlin, J.C., Perloff, R., Sternberg, R.J., and Urbina, S. (1996). Intelligence: knowns and unknowns. *American Psychologist*, 51, 77–101.

Nettlebeck, T. (1987). Inspection time and intelligence. In P.A. Vernon (ed.) *Speed of Information Processing and Intelligence*. New York: Ablex.

Ogbu, J. (1986). The consequences of the American caste sytem. In U. Neisser (ed.) *The School Achievement of Minority Children: New Perspectives*. Hillsdale, NJ: Erlbaum.

Pickens, E. (1970). *Eugenics and the Progressives*. New York: Vanderbilt University Press.

Plomin, R. and Thompson, L.A. (1993). Genetics and high cognitive ability. In *CIBA Foundation Symposium 178: The Origins and Development of High Ability*. Chichester: Wiley.

Quicke, J.C. (1982). *The Cautious Expert*. Milton Keynes: Open University Press.

Raven, J., Raven, J.C., and Court, J.H. (1993). *Manual for Raven's Progressive Matrices and Vocabulary Scales: Section 1*. Oxford: Oxford Psychologists Press.

Richardson, K. (1991a). *Understanding Intelligence*. Buckingham: Open University Press.

Richardson, K. (1991b). Reasoning with Raven – in and out of context. *British Journal of Educational Psychology*, 61, 129–138.

Richardson, K. and Webster, D. (1996a). Analogical reasoning in context: a research note. *British Journal of Educational Psychology*, 66, 23–32.

Richardson, K. and Webster, D.W. (1996b). Recognition of objects from point-light stimuli: evidence of covariation structures in conceptual representation. *British Journal of Psychology*, 87, 1–26.

Robinson, D.N. (1981). *An Intellectual History of Psychology*. New York: Macmillan.

Rogoff, B. and Chavajay, P. (1995). What's become of research on the cultural basis of cognitive development? *American Psychologist*, 50, 859–877.

Scarr, S. (1989). Protecting general intelligence: constructs and consequences for intervention. In R.L. Linn (ed.) *Intelligence: Measurement, Theory and Public Policy*. Urbana: University of Illinois Press.

Scribner, S. (1997). Knowledge at work. In E. Tobach, L.M.W. Martin, R.J. Falmagne, A.S. Scribner, and M.B. Parlee (eds) *Mind and Social Practice: Selected Writings of Sylvia Scribner*. Cambridge: Cambridge University Press.

Spearman, C. (1923). *The Nature of 'Intelligence' and the Principles of Cognition*. London: Macmillan.

Sternberg, R.J. (1984). Towards a triarchic theory of intelligence. *Behavioral and Brain Sciences*, 7, 269–315.

Sternberg, R.J. and Berg, C.A. (1986). Quantitative integration: definitions of intelligence: a comparison of the 1921 and 1986 symposia. In D.K. Detterman and R.J. Sternberg (eds) *What is Intelligence? Contemporary Viewpoints on its Nature and Definition*. Norwood, NJ: Ablex.

Terman, L.M. (1916). *The measurement of intelligence*. Boston: Houghton-Mifflin.

Terman, L.M. (1917). Feeble-minded children in the public schools of California. *School of Society*, 5, 161–165.

Terman, L.M. (1942). The revision procedures. In Q. McNemar (ed.) *The Revision of the Stanford–Binet Scale*. Boston: Houghton-Mifflin.

Terman, L.M. and Merrill, M.A. (1961). *Stanford–Binet Intelligence Scale. Manual for the Third Revision, Form L-M*. Boston: Houghton-Mifflin.

Thorndike, R.L. and Hagen, E.P. (1969). *Measurement and Evaluation in Psychology and Education*. New York: Wiley.

Wagner, R.K. (1994). Context counts: the case of cognitive ability testing for job selection. In R.J. Sternberg and R.K. Wagner (eds) *Mind in Context: Interactionist Perspectives on Human Intelligence*. Cambridge: Cambridge University Press.

Wattenmaker, W.D. (1996). Knowledge structures and linear separability: integrating information in object and social categorisation. *Cognitive Psychology*, 28, 274–328.

Wechsler, D. (1958). *The Measurement and Appraisal of Adult Intelligence*. Baltimore: Williams & Wilkins.

Wolf, T.H. (1973). *Alfred Binet*. Chicago: University of Chicago Press.

5

GENETIC INEQUALITY: HOW DO WE KNOW?

Introduction

The belief in innate inequalities in cognitive ability has long been with us and scholars have long urged recognition of its consequences for social policy and the survival of the state. As indicated in Chapter 1, Plato, in Ancient Greece, warned citizens that only children of 'Gold' can become true Guardians. Shortly after, Cicero was warning the Romans:

> Since the most powerful influence in the choice of a career is exerted by Nature, and the next most powerful by Fortune, we must of course take account of them both in deciding upon our calling in life; but of the two Nature claims the more attention.
>
> (Cicero 1975: 73).

About a hundred years ago Galton tried to prove scientifically the biological basis of social inequality by suggesting the use of twins in conjunction with tests of ability (which subsequently failed his purposes). In the twentieth century there has been a steady flow of claims offering the same message, especially since the founding of compulsory education systems:

> Any recent attempt to base our educational policy for the future on the assumption that there are no . . . important differences, between the average intelligence of the different social classes . . . is likely to be fraught with disastrous consequences for the welfare of the nation as a whole. . . . The facts of genetic inequality are something that we cannot escape.
>
> (Burt 1959: 28)

> More is known about the origins of individual differences in cognitive abilities than any other behavioral dimension. . . . It is clear that genetics play a major role.
>
> (Plomin and Thompson 1993: 77)

And these views have permeated the consciousness of the general public and academic psychologists alike:

> Since the time of Francis Galton, the nature–nurture controversy has been virtually synonymous with the debate on whether genetic factors influence IQ . . . it appears that the issue has been resolved. Snyderman and Rothman (1990) recently surveyed psychologists' beliefs about intelligence and reported that over 90 per cent of those responding agreed that IQ was, at least in part, heritable.
>
> (McGue *et al.* 1993: 59)

Most such claims have been about social classes. But occasionally they have bubbled over into speculations about 'race' differences, as with the furore surrounding the writings of Herrnstein and Murray (1994), and a recent scandal in which a book on IQ had to be withdrawn by the publishers on the eve of publication because of fears of incitement to racial hatred. More generally, though, most psychologists today see themselves in a far more benign, if not, indeed, benevolent, role of seeking genetic knowledge relevant to cognitive abilities, paving the way for practical interventions in the way that dietary interventions have successfully treated certain specific genetic abnormalities such as phenylketonuria.

The main aim of this chapter is to examine the assumptions underlying that quest and the kinds of conclusions that have been reached. Once again I hope to show that the major concepts adopted are as idealistic as the belief in an intractable mental power, or IQ. I also want to demonstrate that in embracing highly idealistic assumptions about genes, environments, and cognitive ability, as the grist of argument, psychologists and behaviour geneticists seem to have entrapped themselves in a web of self-contradictions: for the 'evidence' so frequently presented – in fact, nothing more than simple correlations – actually supports models of genes and environments that they, and anyone else with a modicum of developmental circumspection, simultaneously disavow. Indeed, I think it can be safely said that never before in any field of science have so many arbitrary assumptions been gathered together, *in full knowledge of their invalidity*, as the basis of substantive claims about the nature of people, with so many potentially dire consequences for them.

I examine the evidence on which so much debate rests, and, like many other scrutineers, find it lacking in integrity. This is partly because of the methodological assumptions on which the gathering of such evidence is founded, and which I also show to be weak. But it is the *theoretical* assumptions underlying methods and data interpretations that seem most self-contradictory, and I examine those in some detail. Along the way, I hope to show, once again, that claims like those above have entailed the erection of the 'gene', which is not that known to the developmental geneticist, but a special, idealistic verson, that has a most unlikely existence. This has been particularly the case with recent claims about the discovery of a 'gene for IQ' (see Plomin *et al.* 1994). I hope to show how research pursuing and promoting this idealistic gene has created an unfortunate diversion from the kinds of research we really need to be doing in order to understand better cognitive ability.

How do we know?

Let us start by looking at the quotation from Plomin and Thompson (1993) above. This is a forthright claim about scientific knowledge about cognitive ability, and the nature of differences in it. How can it be so forthright? In Chapter 1, we had to arrive at the honest picture of a 'bewildering variety' of views, of a 'Tower of Babel', and of the 'widespread uncertainty' among scientists about what develops, cognitively, and how it does so, let alone the nature of individual and group differences in it. The view of Plomin and Thompson (1993) – as they go on to explain – is based almost entirely on the use of IQ as a measure of cognitive ability and differences in it. I tried to show in the previous chapter that psychologists have been unable to reach an agreed description of what IQ is, and only cling to it, too, as an idealistic construct.

Even if we *were* to ignore that point, how do we know that genes 'have a major role' in causing *differences* in IQ? After all (apart from single-gene defects with conspicuous effects, affecting a tiny proportion of most human populations), we know nothing about genes related to cognitive ability in normal humans, and thus nothing about who has or has not such genes. The main reason for this is that we would expect that huge numbers of genes would be involved; perhaps hundreds or thousands. Moreover, since we can assume that cognitive ability would have been an evolutionarily important trait in humans, we would have expected any contributory genes to have been integrated into an interactive, flexible, gene system. This would contrast with so-called Mendelian traits which show a direct gene-phenotype association on a one-to-one basis: looking for non-pathological single-gene effects in such systems is akin to looking for the effects of single vehicles on variation in traffic flows in complex motorway systems.

Of course, geneticists have some knowledge about polygenic traits. But almost all meaningful general knowledge has come from breeding experiments in animals, using in-bred (i.e. known) genetic stocks and controlled environments: associations between genetic differences and phenotypic differences among offspring can thus be clearly inferred. In humans we cannot get information this way. So whence the strong claims that IQ variance is a more or less smudged picture of underlying genetic variance?

The answer is that behaviour geneticists have only been able to persevere on this track by making a number of strong assumptions. These assumptions permit the construction of an elementary 'model' of genetic differences/resemblances in humans. This model is then taken to predict certain patterns of differences/resemblances in cognitive ability in certain groups of people (e.g. twins; adopted children). Observations (in fact, almost always IQ measures) are then made to see if such differences/resemblances occur in fact: and thus deciding the extent to which the model and its assumptions are correct.

This is the method that has been widely promulgated by behaviour geneticists and embraced by the majority of psychologists. Now, there are many basic technical problems in the usc of simple 'observational' approaches like this, in which neither genes (nor environments) are actually identified or described, but rather inferences drawn on the basis of assumptions. Strictly speaking, nothing can be proven without controlled experiments (see, e.g. Kempthorne 1978) – but I will not go into that here. Rather, I will stick to the assumptions underlying the 'method' itself. I will be examining these in great detail later. But in order to get a general idea, it is necessary to offer a brief sketch of them first. I can say with some confidence that there are great doubts about them because even their strongest proponents occasionally have to admit that the 'elementary model . . . makes a number of strong assumptions which may not generally be true' (Eaves, Eysenck, and Martin 1986: 48).

This is what we have to assume in adopting the method and data interpretation on which beliefs like those quoted above are based.

1 Cognitive ability can be treated like a simple 'quantitative' physical character (e.g. height or foot size) as far as understanding the causes of its development are concerned.
2 As a polygenic character cognitive ability displays random genetic variability within populations and thus varies across individuals according to the 'normal' or bell-shaped curve.
3 It can be measured and expressed quantitatively by IQ tests.
4 These variable genes can be treated as independent 'charges' which can be added together to arrive at an overall 'genetic charge' for each individual (commonly known as the 'independent and additive' assumption). Thus Plomin *et al.* (1994: 110) can suggest that 'high IQ will develop only if an individual has most of the positive alleles and few of the negative alleles for high IQ'.
5 Likewise, the environment is considered to consist of independent charges or 'factors' whose effects can be added together to determine the overall 'environmental charge'.

6 The phenotypic value (e.g. IQ score) is then just the sum of these gene 'charges' and environmental 'charges' – i.e. according to the 'summation of forces' model, described in Chapter 3.
7 Thus, we have to pretend that there are no 'gene–gene' or 'gene–environment' (or, indeed, 'environment–environment') interactions – i.e. each gene 'charge' will make pretty much the same contribution to the trait whatever other genes are present, whatever the environmental 'charges' in which it operates, and vice versa.
8 Without knowing anything about who has which genes or who has experienced what environments relevant to cognitive ability we can arrive at an estimate of the 'heritability' of cognitive ability, within a population, which tells us what proportion of individual differences is 'genetically determined'.

Fundamentally, these assumptions point to a simple pattern of genetic resemblance among family relatives compared with unrelated people: i.e. 1.0 for monozygotic (MZ) twins, 0.5 for dizygotic (DZ) twins and full siblings, 0.25 for half siblings, and 0.0 for unrelated people. It is the analysis of how well resemblance/variation in IQ among relatives conforms to this genetic pattern that has come to constitute the human 'behaviour genetic' method. I will now illustrate how the method has been applied, and then discuss the problems with it.

The twin method

The method of comparative resemblances was first applied to twins. This was because it was realised long ago that identical twins will have identical genes underlying *all* their traits. Since such twins arise from the splitting of a single egg with the same chromosomes, no one can disagree about that (they are also called monozygotic, MZ or single-egg twins – see Figure 5.1). It can be argued that the similarity among MZ twins is a direct consequence of the similarity of their genes, and thus a measure of the 'heritability' of the trait in question.

The task is then to see to what extent the twin resemblances in IQ match the assumed genetic resemblances. A useful measure of 'intrapair' association is the well-known 'correlation coefficient': perfect association of pairs of scores yields a coefficient of 1.0; no association yields a coefficient of 0.0; a perfectly converse association gives a coefficient of −1.0; with all possible values in between (note that I use the term 'association': a common mistake is to take the correlation coefficient as a measure of 'similarity' of pairs, a point to which I shall return below).

Various formulae have been devised to convert such associations between pairs of relatives into a 'heritability' estimate. Heritability is defined precisely as the proportion of individual differences that can be

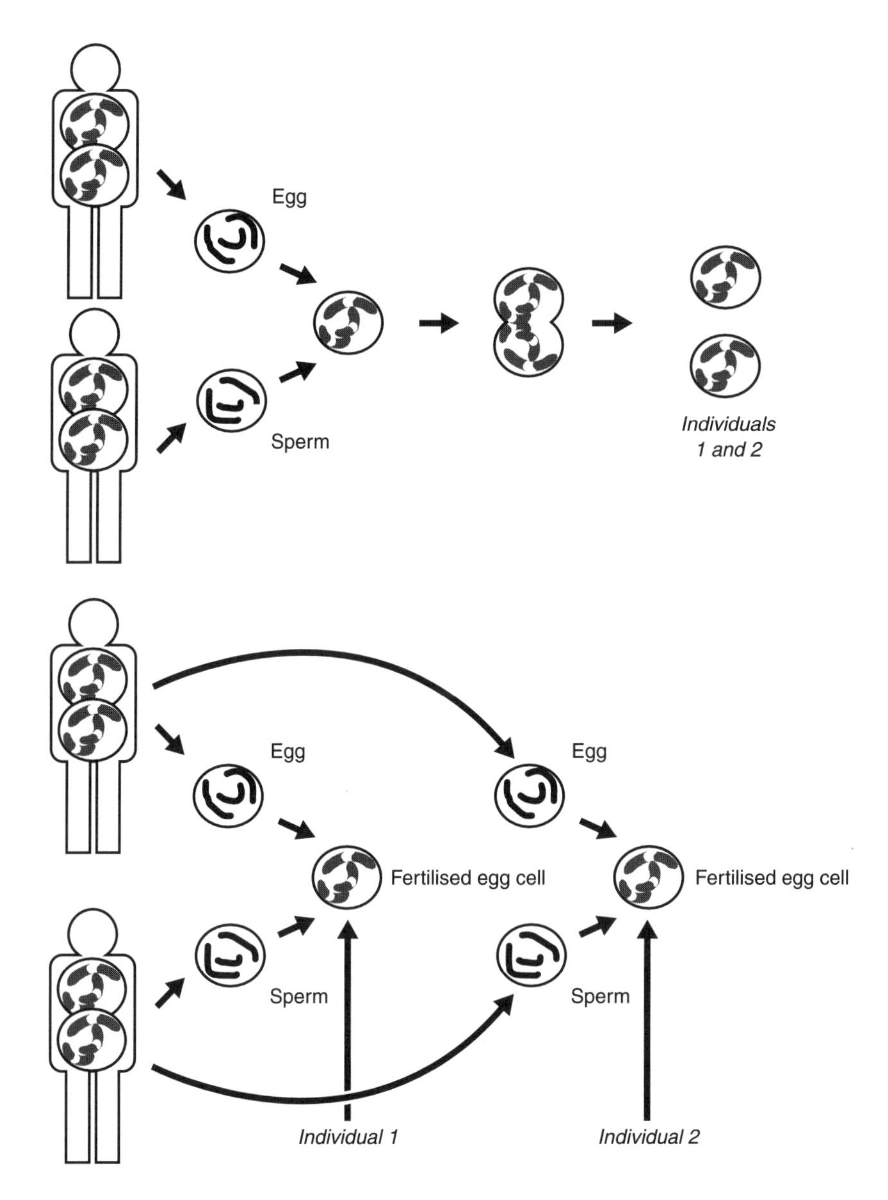

Figure 5.1 The derivation of identical twins from a single egg (top), compared with ordinary siblings and non-identical twins (bottom).

accounted for by genetic differences, and is expressed on a scale from 0 to 1. For example, a heritability of 0.5 indicates that about half of the variability in scores obtained is due to genetic differences in the population being assessed. Note again that, in the whole approach, we do not identify any 'environment' and we do not identify or measure any 'genes', only 'anonymous components of variance' (Plomin and Thompson 1993: 80).

This sounds straightforward enough, but the snag was soon realised: according to the simple 'summary of forces' model (Chapter 1), pairs of identical twins may be similar in a trait because they share related environments, not just identical genes: the pattern of IQ scores could be explained entirely by that. A solution to this problem seemed to present itself in the form of pairs of identical twins who had been reared apart (same genes, different environments). This seemed to be 'controlling' for environmental effects. According to this reasoning, if the mental abilities of such pairs are still associated, this can only be attributed to shared genes and not to shared environments.

Indeed, a number of such studies have been done for IQ, generally showing high degrees of intra-pair association. Plomin and Loehlin (1989) suggest a heritability of over 0.7 based on data summarised by Bouchard and McGue (1981) – meaning that over 70 per cent of the variation in IQ is due to genetic variation. According to Bouchard *et al.* (1990) studies of separated MZ twins suggest that genetic variation accounts for about 70 per cent of the variance in IQ.

Such figures have been extremely influential. Scarr (1993: 10), in a lecture given as President of the American Psychological Association, says that 'The remarkable studies of MZ twins reared in different families challenge many cherished beliefs in developmental psychology'. Plomin and Thompson (1993: 69) refers to 'the powerful design of comparing twins reared apart' suggesting 'a heritability for *g* of 80%'. However, whether or not the theoretical assumptions are correct, there are also a number of other *methodological* assumptions about which there has been constant debate. Let us look at these methodological assumptions first.

Methodological assumption 1: reared apart twins?

Fundamentally, as already indicated, the twin method assumes that individuals are simply sets of independent gene 'charges' immersed, so to speak, in an unreactive, unconscious, world of independent environmental factors – a kind of genetic 'beanbag' suspended in an environmental 'soup' – so that the mere fact of separation approximates a realistic natural experiment with directly interpretable results. Many have pointed to the naivety of this whole approach (e.g. Kempthorne 1978). But even under this assumption, the method demands that twins are separated at birth or very soon afterwards, and into truly separate (in theory they should be 'randomly sampled') environments. Separated identical twins meeting such condition are extremely rare, so studies large enough to meet statistical tests have been at a premium, because they would obviously seem to very convincing.

One such study appeared to be that of Sir Cyril Burt. He studied pairs of identical twins 'reared apart' and cited a correlation in their IQs of 0.771. Between 1943 and 1963 Burt reported correlations on such twins as he accumulated new cases – fifteen pairs at first, fifty-three in the end – with exactly the same correlation of 0.771 on each occasion. It was not until the mid-1970s that this remarkable coincidence was noted and Burt's figures scrutinised. Many, at least, were suspected to have been fabricated, and a long and heated fraud scandal ensued (Kamin 1974; Hearnshaw 1979; for recent appraisals see contributions in Mackintosh 1995).

Four or five other studies are reported regularly in the literature. The analyses of Kamin (1981) and of Rose, Kamin, and Lewontin (1984) of the largest of these (Shields 1962), may be indicative of the general status of these data. Of the forty pairs studied, it turned out that twenty-seven pairs had simply been brought up in different branches of the same family. Usually one child was brought up by the mother and the other by a grandmother, an aunt or a family friend, and wide differences were found in the intra-pair correlations according to the degree of separation.

Moreover, the requirement that the children had been separated at birth and remained separated was rarely met, if at all. Biographical sketches give accounts of twins who were actually separated well after birth (as late as eight years old or later), or only temporarily, before coming together again, who were actually reared by the natural parents who had separated, lived next door to each other, or close by, played together, sat together in school, and so on. That these methodological deviations are important is very clear. Kamin's (1974) analysis of Shield's data showed that the twenty-seven pairs who were raised in different branches of the same family yielded an intra-pair correlation in IQ of 0.83. For thirteen pairs raised in separate families, the correlation was 0.51. Most of those, however, had attended the same school. Among the ten pairs who had shared neither family nor school, the correlation was 0.47. Somewhat similarly, Bronfenbrenner's (1986) analyses of 'separated' identical twin studies showed that correlations among the many pairs who were simply brought up in similar communities was 0.86, while the correlation for pairs in different communities was 0.26.

Shield's study is seen as the best because most other studies have involved very small samples. As Kamin (1974) showed, though, there are just as many doubts among them over whether or not the twins were really separated. Some investigators have attempted to reanalyse the pre-1980s studies by sorting out, according to case studies available, the twins who really were separated from those who were not. This has produced a confusing array of heritability estimates from around zero to 0.8 (reviewed in Locurto 1991).

But these problems continue in more recent studies. For example, in one study frequently referred to as a study of 'reared-apart twins', we find that 'the twins reared apart were separated by the age of 11' (Pedersen *et al.* 1992: 347)! Perhaps the most influential recent study, though, is that of Bouchard *et al.* (1990) published with some prominence in the journal *Science*. Because it illustrates so many of the problems inherent to such studies it is worth dwelling on it a little.

Bouchard *et al.* studied fifty-six pairs of MZ twins 'separated in infancy and reared apart'. The twins were tested as adults on both the Wechsler Adult Intelligence Scales (WAIS) and Raven's Matrices. The intra-pair correlation (mean of both tests) was reported to be 0.75, thus leading to the heritability estimate of around this value. Rapid reading of the report might suggest an almost immaculate study with decisive results. It is not difficult to show, though, that the study is far from a meticulously controlled investigation.

First, the circumstances of separation and rearing, and information about them, are far from clear. We are simply told that 'Circumstances of adoption were sometimes informal' (without any indication of numbers of pairs affected). This suggests, as with all such studies, that members of at least some of the pairs were reared in the households of relatives, or friends, or in the same neighbourhood, and so on. And were they separated at birth? We are told (in their Table 1, p. 224) that 'Time together prior to separation' was 0–48.7 months – i.e. members of at least some of the pairs had spent as much as four years together before being separated. At the very least it ought to make us suspicious that the sample consisted of something other than fifty-six pairs of twins randomly separated at birth and reared apart.

Then there is the method of selection of the sample, which (of necessity) was equally informal. Friends, relatives, care-workers involved in adoption, or 'the re-united twins themselves' (numbers not given), having learned of the project, contacted the research centre in Minnesota. Such a procedure may well produce bias, in that there will be an emphasis on twins who look and behave alike to be selected (or select themselves). Pairs who have little affinity with each other, either in their own or other agents' assessments may not have been offered for study. And, of course, the recruitment procedure will tend to favour those who live in closer proximity to one another.

Now let us look more closely at the 'reared apart' claim. As the quotation above implies, some of the pairs had already reunited before the study began, and it seems safe to assume that some of them were living fairly close to each other. Indeed, we are told that some of the pairs 'actually met at intervals during childhood' (225; again, numbers not given). The authors are quite candid about this, and provide us with a table showing (presumably from twins' self-reporting) two remarkable facts. First, that the 'time apart to first reunion' had a range of 0.5 to 64.7 years (the age range of the twins being nineteen to sixty-eight years). Second, that their 'total contact time' had a mean of 112.5 weeks and ranged from 1 to 1233 weeks. In other words, members of at least some pairs had enjoyed as much as twenty-three years in contact with each other.

These details would seem to confirm the suspicion of Rose, Kamin, and Lewontin (1984: 108) that 'The technical use of the word "separated" by scientists of IQ obviously differs from the usage of the same word by ordinary people'. But Bouchard *et al.* (1990) claim that these flaws do not affect their study. Their confidence in this regard seems to derive from an analysis which suggests little correlation between twin similarity (in fact intra-pair IQ difference) and the amount of contact time. This would seem to fly in the face of common sense, as well as previous studies (e.g. Taylor 1980; Rose and Kaprio 1988; cf. Bouchard *et al.* 1990).

The problem is that, without detailed case studies of twins' histories, as well as the social dynamics of separation and reunion, a blanket correlation simply conceals a host of imponderables about this question. For example, some reanalyses of previous twin studies have shown a curvilinear relationship between contact time and twin similarity (Locurto 1991); a correlation only measures linear association. Moreover, it needs to be remembered that contact *time* tells us nothing about the *quality* of such contact. Some pairs who have only been in contact for a relatively short time may enjoy a great deal of intellectual cross-fertilisation, whereas others who have been in contact for years may have enjoyed little. Finally, there is the obvious point that twins who have spent *all* their time together (i.e. reared together) have IQs which correlate significantly higher compared with MZA's (Bouchard and McGue 1981), clearly indicating some role for contact time.

It is possible that, although 'separated', twins still enjoy similar environments, and this will explain similarity in IQ. Bouchard *et al.* (1990) attempted to eliminate this possibility by asking twins to fill in, from memory, questionnaires about facilities in the home (power tools, sailboat, telescope, and so on) and some other aspects of rearing. Ratings of the similarity of these 'environments' was then correlated with similarity in IQ, and a very low value was obtained.

The problem with this supposed reassurance is that no one has yet described satisfactorily what the 'environment' that promotes IQ development (and thus IQ similarity) actually *is*. As described in Chapter 4, much argument can be made about what counts as 'measures' of the environment for IQ, but it seems unlikely that something as tangible as material possessions, or superficial aspects of rearing, will do. As Locurto

(1991) notes, the environments of two groups of children may *appear* to be the same, when the *effective* environments are quite different.

As with all such studies, the Bouchard *et al.* (1990) report displays irritating insufficiencies in basic information (such as numbers who did spend time together as children or contacted the study together). Socio-economic status is given for fathers but not for mothers, which may be more important. Then there are doubts about measures, of IQ as well as other things. Although it is said that twins were tested on the WAIS and the Raven's, some correlations are only given for the former. Moreover, we are told that twins in the overall study have come from the USA, the UK, Australia, Canada, China, New Zealand, Sweden, and Germany, yet no test is standardised for this vast range of countries, and it is not clear from the report which of these twins is included in the current report, nor *where* they were tested. This is very important because an IQ test, being 'standardised' for age and sex differences for a particular country, means that random pairs of individuals would display a covariation in IQ just by being the same age, same sex, or coming from the same part of the world, unless their scores are corrected with reference to norms. As Bouchard *et al.* (1990) admit, the norms for the WAIS they used are very old and restricted. In view of all this it is perhaps not surprising that one exasperated group of correspondents (Beckwith, Geller, and Sarkar) complained to *Science* as follows:

> Some of their data have not been published in a format that permits independent scrutiny. Investigators should indicate the precise nature of being 'reared apart', including, for example, whether it involved only being raised in separate households within the same community . . . [and] whether they were [adopted] by relatives of the biological parents or friends. In the case of adoptions mediated by welfare organizations, investigators should indicate the criteria by which potential adoptive parents were chosen. Moreover, they should consider the reliability of answers given by identical twins who may be under social pressure to be similar. The fact that data were collected only from the relatively few twins who came forward might also bias the data. . . . Given these problems . . . it is imperative that case studies be fully published. In 12 years the Minnesota group have not provided these case studies, and the two apparently relevant articles cited by (them) contain no case studies.
>
> (Beckwith, Geller, and Sarkar 1991: 191)

There seems little doubt, then, that studies of 'separated' twins are universally – and probably intrinsically – flawed. As Bouchard (1983: 181) notes, 'The requirement of randomization in the MZA design means just that – randomization of cases across trait-relevant environments'. But not only is randomisation clearly breached: we do not know what the trait-relevant environments are (due largely to the fact we are not even sure what the trait is).

In addition, each and every study conceals a host of assumptions about causes, variables, measures, linearity of effects, and the ubiquitous correlation coefficient that cannot be checked. One of the most remarkable general assumptions, indeed, is that ideal studies have been conducted in which the crucial variable (degree of genetic resemblance), manipulated for us by nature, is randomly placed in a kind of bland environmental 'soup' of independent factors, allowing direct and obvious interpretation of results in outcomes, for which we have ideal measures and clear descriptions. There are, of course, flaws in all of these assumptions. Since the studies to date have shared most of those flaws it is scarcely surprising that they have often (but not always) produced similar results.

Yet these data for 'separated twins' are still regularly used as evidence for a heritability estimate, often with a finality of conviction that the results are beyond doubt and beyond debate (Rutter, Simonoff, and Silberg 1993). This conviction has no doubt been encouraged by the trend in recent years of simply turning a blind eye to separate flaws by lumping all the results together as an aggregate set of data. The mean of the separate correlations is then taken as an estimate of the 'truth'. This increases the impressiveness of the numbers and permits the use of sophisticated 'model-fitting' statistical programmes, as if this was overcoming problems instead of compounding them (e.g. McGue *et al.* 1993). It is difficult to imagine this happening around sensitive issues in any other discipline. Yet the practice appears to be condoned by the group set up by the American Psychological Association (the APA group – Neisser *et al.* 1996) to summarise 'knowns and unknowns' in the debate, thus perpetuating the impression among naive readers that they we have ideal data from ideal studies. It seems more reasonable to conclude, as does Locurto (1991: 119), that no study of separated MZ twins exists 'that is neither tainted by fraud nor susceptible to endless reinterpretation'.

Comparison of MZ and DZ twins

Because of these difficulties, investigators have suggested that comparing the resemblance of MZ twins with that of same-sex, non-identical (dizygotic, DZ, or two-egg) twins could be used as a measure of heritability. While MZ twins share all their genes for intelligence, this will not be so for other family relatives, under the assumptions listed above. For example, if the genes associated with IQ are variable in a population, non-identical twins, like ordinary siblings, will only share half of those varying genes, on average (because they receive one assortment from each parent). Using this approach, Plomin and Loehlin (1989) suggest a heritability of 0.52, again based on MZ and DZ twin correlations summarised by Bouchard and McGue (1981).

But this approach is not without problems either. The reason that investigators' first preference is for *separated* MZ twins is the suspicion that, within families, they could be 'treated' alike. In the case of MZ–DZ comparisons the worry is that MZ twins will be treated *more* alike than DZ pairs, so that differences in resemblance in IQ could be due to that. Rejecting this possibility has involved making what is called the 'equal environments assumption'.

Indeed, studies, even of superficial environmental aspects of the home, find them to be *markedly* more similar for MZ twins than for DZ twins. For example, parents treat MZ twins more similarly, the twins are more likely to dress alike, share bedrooms, friends, activities, and so on. Moreover, they are so treated by parents simply because of their degree of similarity in appearance, rather than any knowledge of their exact zygosity (i.e. whether they are MZ or DZ twins): 'Most probably, identical twins are treated more alike because they look and act more alike – at least we found little evidence that parental beliefs about zygosity were important *per se*' (Loehlin and Nichols 1976: 87).

In other words, we would expect such treatments to have differential effects on cognitive resemblances between pairs of twins, and thus render the method useless. What investigators now argue is that this does not matter anyway, because the differences in environmental experience are *not* actually those which affect IQ. 'This possible confounding effect has been examined and, in research to date, does not appear to represent a major problem for the twin design' (Plomin and Daniels 1987: 3).

Methodological assumption 2: the 'equal environments assumption' (EEA)

This is, of course, a crucial issue and we would expect it only to be satisfactorily answered by the highest empirical standards. Surprising as it may seem, nobody really knows what environments affect IQ (in turn, owing to the fact that there is little agreement about what IQ actually *is*). Instead of analysing environments for their causal effects on IQ, again the recourse has been simply to look for correlations between speculative indices of environmental similarity and resemblance in IQ. The assumption has been that, if none can be found, then they do not exist and the method is sound.

In fact only three such studies have claimed to examine the equal environments assumption in relation to 'cognitive' measures. The one most regularly quoted in defence of the assumption is that of Loehlin and Nichols (1976). Strangely enough, this study was not about IQ or cognitive abilities as such, but was mainly a study of *personality* resemblances in twins. The twins were compared on the basis of the (American) National Merit Scholarship Qualifying Test, a test given routinely to high school juniors in the US for educational selection purposes, and which they had already taken. (As we shall see several times in such studies, almost anything that smacks of a 'mental' test is taken to be a surrogate of level of general cognitive ability.)

Those authors concluded as follows:

> Our data have thus not yielded any final and conclusive answer to the heredity–environment question for personality, ability and interests. The data are generally consistent with a substantial influence of the genes in accounting for individual differences in these domains, but they imply a substantial influence of the environment as well – indeed they do not altogether exclude a completely environmentalist position. The data do, it seems to us, have something important to say concerning this environment. And the upshot of what they say is that it operates in remarkably mysterious ways.
>
> (Loehlin and Nichols 1976: 94)

This is what we might naturally expect in studies which are rather in the dark with regard to what those 'abilities' are, and the nature of the 'environment' believed to influence them.

The second much-quoted study said to support the 'equal environments assumption' for cognitive abilities is that reported by Scarr and Carter-Saltzman (1979). These investigators examined the IQs of MZ and DZ twins, and assessed whether their facial and physical resemblance as perceived by others was associated with resemblance in IQ. Twin's resemblances in appearance were estimated from ratings of photographs by eight independent psychologists. The children (ten- to sixteen-year-olds) were tested for their cognitive ability by the well-known Raven's Matrices test. It was found that 'The Ravens Matrices score differences were significantly predicted by differences in appearance as rated by eight psychologists' (Scarr and Carter-Saltzman 1979: 535). The correlation was about 0.27, which may not sound much but could be enough to account totally for MZ–DZ differences (see below, p. 153). In other words, this result does *not* confirm the equal environments assumption in the way that is frequently claimed. This correlation was only reduced to a less significant value by averaging the Raven's scores with those from a less well-known, and little used, ability test. This combined value seems to be the one that proponents of the method have seized upon in support of the EEA. But another result (and one which illustrates the erratic nature of results in this area) was that, 'For the DZs, larger *differences* in Raven scores were significantly related to greater similarity in appearance, as rated by eight psychologists' (Scarr and Carter-Saltzman 1979: 536, emphasis added).

This result suggests complex social dynamics in the nature of experience associated with IQ. Why should DZ twins who look more alike have greater differences in IQ? We have already seen that 'the environment works in mysterious ways'. Perhaps, as Loehlin and Nichols (1976: 93) noted from their own findings, some twin pairs (e.g. those who believe they are *not* identical twins) actually contrive to make themselves *different* from each other through what they refer to as 'effects of competition or contrast between members of twin pairs'. Again, this sounds like a negation of the equal environments assumption: such factors alone could easily account for MZ–DZ differences in resemblances. Either way, we can hardly conclude from such confused findings that 'Perceived similarity is not an important bias in studies of genetic variance in intellectual skills' (Scarr and Carter-Saltzman 1979: 541).

A third study is that of Matheny, Wilson, and Dolan (1976). They used 121 MZ twin pairs and 70 DZs that had appeared in a large longitudinal study of child development in the USA. Scores of physical similarity were obtained by sending a questionnaire to parents containing ten items on physical resemblance, including facial appearance, height, whether or not the twins were ever mistaken for one another, and so on. The composite scores that resulted were then correlated with intra-pair differences in intelligence test scores (the Stanford–Binet or the WISC). No significant association was found.

This appears to support the equal environments assumption in that differences in appearance (and thus of presumptive treatment in the home) did not appear to affect differences in IQ. Yet this, too, is far from being a perfect study. The small numbers of DZ twins in comparison to the numbers of MZ twins (seventy compared with 121 pairs) suggest a large measure of selection in the sample (perhaps from DZ parents being non-responders, or that there was a tendency only for the parents of twins, MZ as well as DZ, who were most similar to respond).

In addition, there may be problems stemming from the peculiar structure of the physical similarity scale. On five of the criteria used (e.g. height, hair colour, facial appearance), parents were asked to respond with '0' (not at all similar), '1' (somewhat similar), or '2' (exactly similar). On another five criteria (e.g. 'is it hard for strangers to tell them apart?'), only a dichotomous scale was used: 0 ('no') or 2 ('yes'). Responses were then added together to make a twenty-point scale of twin similarity. However, it is difficult to imagine parents of identical twins responding with zero to many of these questions, or parents of DZ twins responding with many '2s', and scores will therefore tend to 'confirm' one another (a '2' on one question strongly tending to be accompanied by a '2' on all others). In consequence, the scores will bunch towards upper and lower ends of the scale for MZ and DZ twins respectively, and any association with IQ similarity will tend to be strongly curvilinear. A correlation only measures *linear* association.

Another study occasionally referred to in support of the EEA for cognitive ability is that of Goodman and Stevenson (1989). This suggested that 'MZ twins had strikingly similar intraclass correlations on our objective measures of intelligence' (Goodman and Stevenson: 695) whether they were *recognised* as identical twins by parents or not. In other words, differences in MZ recognition, presumably related to degrees of resemblance of appearance, did not seem to elicit differences in treatments from parents in such a way as to affect resemblances in IQ.

It needs to be pointed out, however, that the 'objective intelligence measure' was a little unusual, to say the least. It consisted of scores on only *three* (out of eighty!) items of the WISC. Moreover, these were not originally used in the study as measures of intelligence at all, but as measures of 'freedom from distraction' in a study of hyperactivity (for example, one of the three was an item in which children have to put a line across as many squares as possible in a given time). This would seem to exemplify the adaptable 'make do' empirical culture that has, unfortunately, come to surround such arguments.

Note that studies of this kind only assess within-family treatment affects. Physical similarity/dissimilarity is much more likely to lead to treatment effects by teachers, peers, and others than by parents and siblings – parents even of identical twins are able to tell them apart most of the time. And treatment effects may vary across cultures. Indeed, in a study of black (African-American) twins and white American twins, Scarr (1981) found that the patterns of correlations among MZ and DZ twins differed significantly between the two groups. This could only be explained by different patterns of treatment effects across the two ethnic groups.

In sum, these studies illustrate the many imponderables which accompany uncontrolled observational studies attempting to answer a question which, strictly speaking, can only be answered by controlled experimental studies (which, in turn, would require far clearer hypotheses about what IQ measures, and precisely what aspects of the environment affect it, and how). Far from inviting scepticism, however, the EEA issue, among behaviour geneticists, is now considered to have been settled. Consequently, journals have been increasingly filled with comparisons of MZ and DZ twin pairs being converted directly into heritability estimates with no regard at all for the possibility that the differences can be anything other than 'genetic'. Again, a common strategy is for the results of flawed single studies to be simply rolled up together, as if the impressiveness of numbers somehow eliminates the common flaws. In my view, such strategies do not meet the empirical standards required in this area.

The adopted child method

Because of these uncertainties over twins, another approach has been to study the IQs of adopted children in comparison with that of their natural and their adoptive parents.

> Adopted children . . . provide almost as useful data as the rare identical twins reared apart. Adopted children are not genetically descended from the family of rearing, and therefore environmental differences between families are not confounded with genetic differences in the children if the adopted children are randomly placed by adoption agencies.
>
> (Scarr and Carter-Saltzman 1982: 833)

This approach, then, depends on comparing the resemblances between children and their biological parents (with whom, it is reasoned, they share half of any variable genes for intelligence, on average) with that between the same children and their adoptive parents (with whom they share none of their genes for intelligence, theoretically). Again, formulae exist for converting any differences in resemblance into heritability estimates.

Several such studies have been done over the last fifty years or so and these have revealed that the IQs of adopted children correlate more highly with those of their biological parents than with those of their adoptive parents. This has led to claims of substantial heritability for IQ. For example, one study indicated that the adopted child–natural parent correlation was around 0.35; while the adopted child–adoptive parent correlation was 0.1 or less (Honzik 1957). This suggests a heritability of at least 0.50. These results are taken by Scarr and Carter-Saltzman (1982: 842) to be 'dramatic illustration of the effects of genetic resemblance on intellectual resemblance'. More recent results from the Texas Adoption Study have been taken to suggest heritabilities of 0.62 at age eight. (Horn, Loehlin, and Willerman 1979) and 0.70 at age seventeen (Horn, Loehlin, and Willerman 1986, cited by Plomin and Loehlin 1989). Results from the Colorado Adoption Project have been taken to suggest a heritability of 0.50 at age four (Plomin, DeFries, and Fulker 1988).

Methodological assumptions in the adopted child method

The view that such studies have provided useful data may be even more idealistic than it turned out to be with the so-called 'separated' identical twins. As with twins, a fundamental assumption of adoption studies is that of asocial organisms (the children) of different genetic make-up being implanted into essentially asocial, non-dynamic environments where we can clearly tell what are the effects of the genes and what are the effects of the environments. I will call this the 'social vacuum' assumption.

Even within the 'social vacuum' paradigm, though, there are several minimum requirements for an adoptive child study aiming to assess heritability of IQ. First, the adopted children need to be separated from their natural parents at or around birth. Second, we need a measure of the natural parents' IQ (as a surrogate for what the child's IQ 'might have been'; Locurto 1991). Then we need, of course, IQ measurements for the adoptive parents (ideally the mid-point of *both* parents). Finally, we need to know that adoptive children have been allocated to their adoptive family environments randomly.

Even these simple methodological requirements have never been met. For instance, although a crucial correlation is that between adopted child and natural parents, in only two studies have the latter's IQs actually been measured. In one of these (Horn, Loehlin, and Willerman 1979) very little difference was actually found in the crucial correlation contrasts. In all other studies the practice has been to use a 'proxy', usually the mothers' level of education, as a 'measure' of IQ. Similar surrogate measures have regularly been used for the adoptive parents.

But the social vacuum assumption becomes confounded by complex social dynamics right at the start of adoption: it is now widely recognised that 'selective placement' of children for adoption often takes place by adopting agencies. From their knowledge of the character of the biological parents, these agencies tend to have preconceived ideas about the personality and intelligence that the child is going to have, and place it with what they think will be a compatible family environment (Scarr 1992). The effects of this genetic prescience (itself, perhaps, an ironic consequence of views on genetic determination of intelligence published by psychologists), are ones we can only speculate about. But it seems highly unlikely that parent–child correlations will be entirely unaffected.

In addition, adoptive parents are likely to be socially quite different from parents who give up their children in many ways which could affect the adopted child–adoptive parent correlations. They obviously badly want children, they are carefully screened as prospective parents, they tend to be much older, much better off, and of higher social status. They are thus likely to present a far more uniform range of experience for adopted children than the biological parents might have done. The effect of this may be to suppress the adopted child–adoptive parent correlation compared with the adopted child–natural parent correlation (Rose, Kamin, and Lewontin 1984), but there are a number of other possibilities.

Various social-psychological confounds are likely to suppress adopted child–adoptive parent correlations. As Jackson notes, the adopted child method

> is engaging because of its simplicity. However, it is critically incomplete as a model of what actually happens in family life when an adoption is involved. . . . All adoption studies that follow the human behavioral genetics model fail to account for adoptive status as a highly important, culturally imposed social factor in the psychological environment of development.
>
> (Jackson 1993: 1323)

Jackson also notes that it is well known that adoptive parents treat their adopted children differently from their normal children, *because they are adopted*: they may even go out of their way to make them different, to 'allow' the child 'freer' development, and have less clear hopes and aspirations for them. Clinicians who work with adoptive families report how adoptive parents, from the moment of adoption, worry about the personalities, bloodlines, and social histories of the natural parents of their adopted child, and this affects the parent–adopted child relationship (Hajal and Rosenberg 1991; cf. Jackson 1993). The fact that adoptive parents 'hold strong beliefs in the influence of heredity' (Terwogt, Hoeksma, and Koops 1993: 499), compared with ordinary parents, may also have complex psychological consequences for their adoptive children's development.

Later, adolescent adoptees can become highly conscious of their special identity and react to adoptive parents' standards and values, thus reducing further any psychological similarity (Brodzinsky, Singer, and Braff 1984; cf. Baumrind 1993). Adopted children are more likely to be different in physical appearance from their adoptive parents compared with their natural parents (and we know that physical appearances do affect how children are treated). It is reported that children who are aware of such physical differences, and that this can be attributed to their adopted status, can feel at a disadvantage and have lower self-esteem because of it (Rosenberg and Horner 1991; cf. Jackson 1993).

To feel that we are somehow getting pure, unadulterated correlations, whose meaning can be taken at face value, from such a dynamic social system, seems risky, to say the least. Some attempts have been made to overcome the more obvious confounds by introducing 'corrections' to the data, though these largely consist of guess-work. For example, Scarr and Carter-Saltzman (1982: 854–855), after estimating adoptive parents' IQs from their level of education, corrected these estimates further for 'restriction of range', '[b]ecause the adoptive parents are quite bright'. Then they corrected for the effects of selective placement by arbitrarily subtracting 'half of the selective placement coefficient' (a 'guesstimate' of the size of the effect of selective placement) from the correlations in question. They conclude that 'Other corrections could be justified by the data set . . . . [This] simple figuring of these data yields heritabilities of 0.4 to 0.7'!

Unsurprisingly, in view of investigators' disregard for the social dynamics of the adopted status, there has been much puzzlement over other correlations within adopted families. For example, Scarr and Carter-Saltzman (1982) note from their adoption study how adopted children, in spite of lack of correlation with adoptive parents' IQ, often come to resemble the IQs of other siblings in the adoptive family just as much as do *any* siblings. After commenting that 'the heritabilities calculated from young sibling data are drastically different from those calculated from the parent–child data', and tracing how this puzzling picture has emerged in other studies, they conclude, 'This study only makes the picture worse' (Scarr and Carter-Saltzman: 857). Locurto (1991) in a brief review cites correlations between adoptive children and other children in the same family as large as 0.49 and as small as −0.16 (i.e. negatively associated). And, as already mentioned, correlations found with younger children change drastically in adolescence. As with twin studies, in other words, the raw results are bewilderingly varied.

One reliable finding from adoption studies, however, is that the IQs of adopted children usually increase dramatically after adoption, compared with IQs of their natural parents – up to sixteen points in one French study (Schiff and Lewontin 1986; Locurto 1991). What this means is that adoptive children become far more *like* their adoptive parents in IQ than their natural parents, even though correlations alone suggest a different picture. How can this be so? The answer lies in the (often misunderstood) meaning of correlation. The point is that the within-class correlation method used in twin and adoption studies is not a measure of 'similarity' between pairs of subjects (as it is usually presented) but one of *covariation* from the population mean – i.e. the extent to which pairs of scores 'vary together' from the means of their respective populations.

This has significance for both the twin and adopted child methodologies. Thus, it is quite possible for members of twin pairs, or adopted child–natural parent pairs, to diverge widely on IQ scores but still exhibit a large correlation so long as they vary 'together', even very slightly. For example, of the following pairs of scores, only one of the members on the left is 'similar' to any of the corresponding members on the right, yet the pairs show a perfect correlation of 1.0 because the pairs vary together from the mean.

80	98
90	99
100	100
110	101
120	102

To illustrate the effect in adoption studies, imagine that the middle column of the figures below are the IQ scores of a group of adopted children, while those to the left are from their natural parents and those to the right are from their adoptive parents. The children are clearly more similar to their adoptive parents in IQ than to their natural parents (mean difference of 10 compared with 32); yet their IQ scores *correlate* more highly with those of natural parents than with adoptive parents (around 0.72 compared with 0.45).

Natural parent	*Adopted child*	*Adoptive parent*
90	110	120
85	125	140
98	112	122
87	127	122
105	133	126
100	135	130
112	145	128
110	147	132

In fact, in adopted child studies we are talking about very small correlations of the order of 0.2 to 0.3, with *either* set of parents, anyway. But the point is that a correlation may appear in *any* pairs of subjects who can be quite dissimilar in the variable in question. This was illustrated in one study of eleven pairs of separated identical twins, which found a correlation of 0.66 between them, even though the average difference between the twins was thirteen points (Taylor 1980) – i.e. almost as big as would be expected for any random pair of children. Clearly any, even small, effect that brings about a covariation among otherwise disparate scores will create such a correlation, and the illusion of closer similarity.

Alternative interpretations of nature–nurture data

Apart from the overall 'genetic beanbag and random soup' assumption, mentioned earlier, the problem dogging all these studies is that any gradation of genetic similarity associated with cognitive ability across families can be closely paralleled by gradation of environmental or cultural similarity. Partitioning of causes cannot be convincingly achieved in such circumstances, even if genes and environments could be conceived as independent additive factors. The fact that we do not know what the relevant environments are (largely because we cannot really agree about what the *character* is) hardly helps. What particularly interests me, though, is the ease and haste with which simple correlations are seized upon and put forward, with the insistence that the they simply *must* be due to genetic variation.

Given the methodological problems we have looked at, such correlations can just as plausibly be accounted for by unsuspected shared environments (or rather shared cultural osmosis). There are many characters for which a pattern of correlations among relatives could be conceived of as reflecting 'genetic variance' but in which variance is entirely and *obviously* cultural: similarity of dress, of hair-style, of sports activities, hobbies, other interests, diet, vocabulary, dialect, syntax, and so on. In such cases gradations of correlations among increasingly distant relatives would most obviously be interpreted, not as genetically determined, but as a reflection of diminishing degree of shared culture. Of course, I argued in Chapter 4 that IQ is just such a cultural measure.

It is possible, or even probable, for there to be an intra-pair correlation in any of these measures (including IQ), even among separated twins (at least those studied so far). One obvious reason for such a possibility is that many or most 'separated' twins have been brought up in different branches of the same family, the same neighbourhood, community, or even just similar cultures. Twins brought up in *any* similarity of culture will acquire similar knowledge resources and the patterns of reasoning that go with it. As argued in Chapter 4, IQ appears to be a 'sampler' of precisely these cultural affiliations, rather than a measure of cognitive power as such. And, as detailed above, there is evidence that correlations among 'separated' twins in IQ vary according to the degree of the shared cultural experience (or at least geographic proximity). Moreover, such intra-pair correlations may well be higher than they are for ordinary siblings because the twins are the same age and the same sex, or because the more alike twins tend to have been brought into the study. Or they will be higher than those of DZ twins because of the latter's *social* reaction to each other.

But even perfectly separated twins may still, in effect, continue to share cultural experience. Twins reside in a dynamic social context, in which other people react to them on the basis of their appearance, special status, and so on, according to culturally engendered beliefs. The effects of such social reactivity is most clearly seen in the way that same-sex pairs of twins correlate significantly higher in IQ than different-sex pairs. If social reactions to appearance can discriminate – and create divergent effects on IQ – on the basis of sex, they can also work the other way and unify the experiences even of separated MZ twins. MZ twins look alike, and how we look sets in train predictable sets of cultural experiences.

We know that such effects do exist. For example, we know that facial and other physical appearance strongly affects how children are treated outside the family, the intelligence they are attributed with, and so on, and that this affects their self-concept (Langlois 1986). A similar social effect of inferences drawn from a physical trait to influence, secondarily, a psychological one has long been known to apply with physical height

(Gillis 1982). It has been shown that such dynamic social 'feedback' to children can have deep and durable consequences for their identity and sense of cognitive self-efficacy (Bandura *et al.* 1996).

Obviously, such an attribution process consists of preconceptions prevalent in a given culture, and created at a socio-cognitive level. Again, trait values emerging from one level of regulations (the genomic or epigenetic) become incorporated in a higher set of regulations (the socio-cognitive). There they form the basis of trait differences that may have *no* genomic basis, except in this indirect sense. Such attribution effects have been shown to act most strongly in the case of gender (Lloyd and Duveen 1991), and a 'transactional' model of the process is suggested in Figure 5.2.

Given that MZ twins will continue to look alike, and adopted children will tend to continue to look more like their natural parents than their adoptive parents, and thus share common social reactions to them, they continue to share part of their cultural environments, even though living apart. Even a very small effect of this sort could explain at least part of the crucial correlations. These effects would be expected to increase with age, and typical correlations seem to bear this out (Pedersen *et al.* 1992).

These are cases of 'indirect' gene–environment interaction. With DZ twins, as we saw from the data of Loehlin and Nichols (1976) and Scarr and Carter-Saltzman (1979), such interactions may work the opposite way: the twins consciously act to make themselves as *different* from each other as possible. This effect may well reduce their correlation in IQ. There may be other kinds of gene–environment interaction. These will create measured variance; but, in the typical analyses, according to the additive and independent assumptions, it will only be interpreted as *genetic* variance (see further on this below, p. 164). The contribution of these interactions to the typical correlations, although highly plausible,

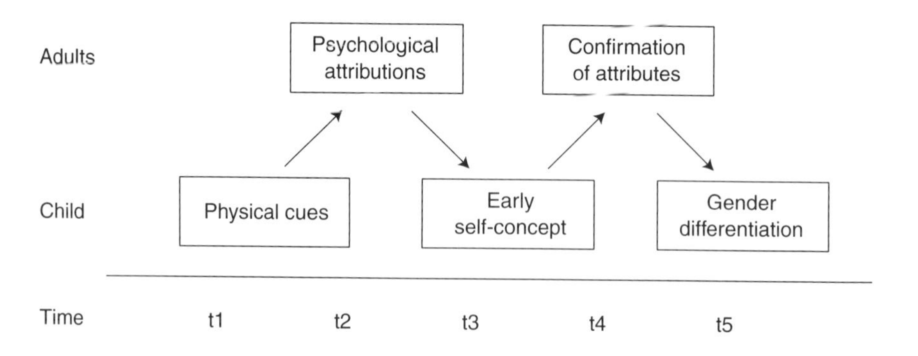

Figure 5.2 A transactional model of gender development.

is entirely unknown. Again, in such contexts, with such measures, it would be surprising if there were not *some* correlations among twins.

None of these possible alternative interpretations of kinship correlations has been seriously entertained because of the dominance of the genetic one. But the most important reason for entertaining them is the sheer implausibility of the 'genetic model' (and its attendant assumptions) actually employed in making genetic interpretations. Even the strongest proponents of the genetic model occasionally admit this implausibility, and this is what I move on to after the following summary.

Summary of methodological assumptions

Should this brief glimpse of the quality of data emerging from behaviour genetic studies lead us to accept the claim of Plomin and others about knowing so much about the genetics of cognitive ability? In so-called 'separated identical twin' studies the 'separation assumption' is most clearly breached, and most investigators most of the time have acknowledged this. It is regrettable that the fact is often forgotten by researchers, including, seemingly, the APA group (Neisser *et al.* 1996). In MZ–DZ twin comparisons, the 'equal environments assumption' is almost completely without justification. This is partly because attempts to examine it have been perfunctory, and partly because we have little, if any, idea of what environmental effects we should be equalising anyway. Finally, the 'social vacuum' assumption, which is the mainstay of the adopted child methodology, appears to be condemned by even a cursory view of the social dynamics of the adoptive child situation.

Instead of 'clean' data, then, we seem to be confronted with a conglomeration of murky correlations, approximations, estimates, and 'corrections', as well as divergent results and *ad hoc* interpretations. Many authors have complained about how this data-collection has entailed relaxing standards of investigation, which, in this of all fields, ought to be of the highest calibre. As Kamin puts it,

> The data are not after all, the product of clearly designed and well-constructed experimentation. They are necessarily correlational data, collected in difficult and inevitably flawed field settings. The patterns discerned within such data are many, and complex. The interpretation of these complex patterns, I believe, must reflect the investigators' theoretical bias.
>
> (Kamin 1981: 468)

And in their survey of the literature Schiff and Lewontin concluded as follows:

> One of the most striking features of the literature and the discussions of human behavioural genetics, and especially of human intelligence, is the degree to which a supposedly 'scientific' field is permeated with basic conceptual and experimental errors. Indeed . . . much of the discussion of the biology of intelligence would simply evaporate if fundamental biological and statistical notions were applied to the genetics of human behaviour with the same degree of rigour and logic that is standard in, say, the study of milk yield in cattle or body weight in mice.
>
> (Schiff and Lewontin 1986: 169)

Of course, I have little doubt that some readers, committed to the standard behaviour genetic assumptions, will suggest that the above account is biased, or has omitted to mention some crucial bit of evidence or other that I am currently unaware of. Yet the 'evidence' surveyed is precisely that which figures most prominently in reviews. It would be nice to be shown *either* how the empirical weaknesses pinpointed are *not* weaknesses, *or* a study that is devoid of them. The plain truth is that there are no such studies.

But the problems run much deeper than that. These methodological problems are more than *practical* difficulties that can be overcome by more and more studies, or pooling or averaging across the inferior ones already done. On the contrary, they arise from deeper theoretical assumptions which I now go on to examine. It appears, in fact, that the 'evidence' outlined above supports a genetic model that many behaviour geneticists themselves – in a self-contradictory way – often disavow. It is as if the correlations presented are giving theorists the right to say 'genes' in one breath, while the model of genes on which such interpretation is based is rejected in the next!

Theoretical assumptions

These assumptions, as mentioned previously, concern the nature of cognitive ability, the nature of genetic variability and gene action, and the nature of the environmental variability and environmental action. It is worth returning to my list outlined earlier (pp. 137–138) to consider these more closely.

Cognitive ability

1 Cognitive ability can be treated like a simple 'quantitative' physical character (e.g. height or foot size) as far as understanding the causes of its development are concerned.
2 As a polygenic character cognitive ability displays random genetic

variability within populations and thus varies across individuals according to the 'normal' or bell-shaped curve.
3 It can be measured and expressed quantitatively by IQ tests.

As mentioned in Chapter 4, the APA group recognises that there is no agreement about what '*g*' is, and our future understanding of IQ may be quite different from what it is today (Neisser *et al.* 1996). This seems a sensible conclusion, given the extent of disagreement among those who seriously ask themselves what IQ is. Bouchard and Propping (1993: 6), for example, point to 'a fundamental division of opinion in the behavioural sciences regarding the importance of *g*, its biological reality, and its validity relative to special abilities'. Now if we want to make claims about the genetics of a function we need to be absolutely clear about what that function is. Yet IQ theorists cannot even agree about whether the test scores they handle reflect a single function, several different functions, or even dozens of different functions. This seems a far from substantial basis for studying the genetics of a function.

Remember also, from Chapter 4, the results of the survey of Detterman and Sternberg (1986) and their analysis of declared attributes of IQ. Twenty-five attributes were mentioned: but only *three* of these were mentioned by 25 per cent or more of respondents. Over a third of the attributes were mentioned by less than 10 per cent of respondents. If we had been talking to potato scientists, we would quickly conclude that they have only a vague idealistic notion of what potatoes are, and are therefore in a rather poor position to be telling the rest of us about the 'genetics' of potatoes.

IQ psychologists persevere by pretending that these problems do not really matter, because IQ is still a character which can be studied just like any other (whether anatomical, physiological, or cognitive). The APA group argues that, 'As far as behaviour genetic methods are concerned, there is nothing unique about psychometric intelligence relative to other traits or abilities' and that 'any reliably measured trait can be analysed by such methods' (Neisser *et al.* 1996: 86). Clearly, such methods *need* to assume that measures like IQ arise from a single 'quantitative' function, exhibiting a normal distribution: only on such grounds can scores be easily broken down, statistically, into components of variation. Again, the standard model is a directly measurable trait like height or head circumference. As we saw in Chapter 4, IQ tests have been routinely constructed to *display* a normal distribution of scores in a population: so we cannot say whether there is a normal distribution in nature or not. However, such distributions are the exception rather than the rule in measurable functions (Micceri 1989).

I have suggested in several places already that such evasions and reductions have played a large part in the perpetuation of nature–nurture debates. Trait values in cognitive abilities are *not* measures of tangible physical features like height or weight. Rather, they are best treated as steady states in a dynamic, self-organising system several levels removed from any variation in the genes.

The Genes

4 These variable genes can be treated as independent 'charges' which can be added together to arrive at an overall 'genetic charge' for each individual (commonly known as the 'independent and additive' assumption). Thus Plomin *et al.* (1994: 110) can suggest that 'high IQ will develop only if an individual has most of the positive alleles and few of the negative alleles for high IQ'.

As mentioned several times, we in no way *discover* genetic variability underlying cognitive ability: psychologists who call themselves behaviour geneticists simply expect such variability to exist and then see if correlations among relatives are consistent with it, usually *without seriously entertaining alternative explanations for those correlations.*

Chapter 3 gave a brief glimpse of the genetics of development, which suggest rather rigid constraints on 'random variation' and 'independent and additive' effects among genes. Taking the assumption of random genetic variation first: genes for evolutionarily important traits do not accumulate as random mixtures like beans in a bag. First, natural selection can quickly reduce genetic variability in characters important to survival, while genetic variability in other characteristics can persist. Second, genes underlying evolved characters tend to become incorporated into cooperative 'teams' in which the effects of variant alleles can be moderated or eliminated altogether.

There are many empirical studies confirming the first of these points. Indeed, it has long been known in experimental agricultural breeding programmes that selection quickly 'uses up' genetic variation within groups (Lerner 1968). Characters which are 'important' can have genetic variability at or close to zero. In over a hundred properly controlled studies in animals it has also been shown that characters important to survival in the wild have tiny heritabilities averaging around 0.1; even evolutionarily less important characters appear to have heritabilities of only around 0.35 (Roff and Mousseau 1987, cited by Barton and Turelli 1989). Indeed, geneticists have often pointed out that characters showing much additive genetic variation are those that have been of lesser importance evolutionarily, and thus likely to be cases of 'genetic junk' or 'evolutionary dross'. The second point, about regulated gene systems will be discussed further below.

These principles suggest a major problem with the assumption of grades of genetic relatedness among relatives and non-relatives. In fact, DZ twins may be as alike genetically, for important traits, as are MZ twins; adopted children, for some traits, may well have a genetic identity with *both* their natural and adoptive parents. This principle seems especially valid when we learn that genomic and other developmental regulations can easily accommodate variation in structural genes with no effect on phenotypes (Rollo 1995). This is, of course, only to restate the fact that changeable environments have encouraged the evolution, not of singly favourable allelic 'charges', but of harmonious gene combinations that act as interactive partners in a regulatory developmental system, sensitive to the world inside and outside the organism. This is a far cry from the simple additive model on which such strong statements about genetic variation in IQ are based.

It is puzzling why investigators who are swift to seek Darwinian underpinning for their ideas ignore one of its key principles in their investigative assumptions – even though they sometimes seem aware of it. Indeed, Plomin (1994: 5, 6) explicitly acknowledges that 'the vast majority of DNA is the same for all members of our species'; that 'evolutionary arguments about the adaptiveness of species-typical characteristics are not irrelevant and they certainly are beguiling'; and that 'characteristics that have been subject to strong directional selection will not show genetic variability because strong selection exhausts genetic variability. In other words, when genetic variability is found among individuals in our species, it is likely that the trait was not important evolutionarily'.

In humans, Plomin (1994) uses the hypothetical example of language.

> If it is the case that our species is a natural language user, this does not imply that differences among individual members of the species in their facility with language are also genetic in origin. Individual differences in language acquisition could be entirely environmental in origin.
>
> (Plomin 1994: 5–6)

The same argument could, of course, also be applied to other evolutionarily important psychological traits such as cognitive ability, of which humans are also 'natural users'.

It seems, in other words, that investigators appear to want to have it both ways, in the sense that they subscribe to evolutionary principles, but then erect a genetic model that may well be contrary to those principles. In sum, grave doubts surround the usual theoretical assumptions about genetic variability underlying evolutionarily important characters (of which we must assume cognitive ability to be one), and which are the crux of the model-fitting approach.

The environment

5 Likewise, the environment is considered to consist of independent charges or 'factors' whose effects can be added together to determine the overall 'environmental charge'.

In a strategy of investigation in which we do not identify any 'environment' and we do not identify any 'genes', only 'anonymous components of variance' (Plomin and Thompson 1993: 80), the environment is simply an anonymous set of forces 'left over' in the total variance after genetic variance has been assessed. Consequently, there is little more to be said about the environment in the 'behaviour-genetic' method, even though, properly analysed in its complex social forms, it could easily give an account of interesting resemblances among people who share environments.

It is important to point out, though, that among psychologists generally there is much uncertainty and disagreement about what 'the environment' relevant to the development of IQ, even theoretically, might be. This is no doubt partly because we have so little agreement about what IQ is, anyway. Mascie-Taylor (1993), on the basis of survey data, shows that variation in IQ is statistically associated with differences in home environment, obstetric complications, maternal smoking in pregnancy, occupational group, gender, number of family and school moves, geographic region, nutritional status, height, and so on, and many of these intercorrelate. Obviously, any of these correlations may conceal complex causal mechanisms or may be indirect or entirely spurious.

Such associations are interesting, but they do not tell us about causes and action. As Bradley points out

> For years, developmental psychologists have disagreed about the nature of environmental action, that is, how a person's experience with the environment influences the course of development. . . . At one extreme environmental action is viewed as holistic: children react to the environment as a composite of many factors. . . . Others see environmental action as highly specific: particular aspects of the environment effect change in particular domains of development at certain critical points in time. Unfortunately the few studies completed do not permit a determination of which view is most plausible.
>
> (Bradley 1994: 256)

Further attempts to identify and measure environmental variables relevant to cognitive development reveal specific assumptions about the *nature* of the environment and how it works. These have been based almost entirely on gross indicators in questionnaire responses about aspects of the home, and impressions gleaned in interviews about quality of parenting. This notion that the 'goodness' of the environment for cognitive development can be measured in terms of scores on a few simple independent parameters is, of course, endemic to the more general view of development as the sum of genetic and environmental 'forces'. But this view has been very widespread, as Wachs (1992) notes.

The contrary view is that the child is in possession of cognitive regulations which are, themselves, creative centres of development, especially when they involve action in social contexts. Many studies of early communication and cognition in the 1970s revealed that the environment of the human, right from the first days of infancy, consists of a structured world of human purposes and intentions, embedded in a cultural framework (e.g. Bruner 1974; Trevarthen and Logotheti 1989). These do not implicate independent environmental 'factors' that can simply be added together to give an environmental score. Rather, like genes, the developmental import of these are conditioned at several interacting levels, in which the interactions themselves are of much greater significance than the independent factors.

For example, an 'environmental' factor such as school exclusion, assumed to be a punishment for bad behaviour, may actually be rewarding to some children, depending on the developmental targets prominent in their cultures. Studies have shown that interventions which overlook this structure have limited success. Thus, children's adjustment to day-care programmes has been shown to have little to do with the quality of the programmes themselves, and much more to do with their mothers' attitudes (Everson, Sarnat, and Amron 1981; cf. Woodhead 1996), which themselves are conditioned by cultural or subcultural perceptions. In this framework, development of cognitive abilities is promoted or otherwise, not because of the 'environment' score (or a lucky or unlucky deal of genes) but by the extent to which experience provides access to the cultural tools of society. Some behaviour geneticists appear to be beginning to appreciate that the environment is a dynamic structure of interacting factors, although, as Plomin (1994: 33) admits, 'few measures of the environment actually reflect this model'. This may be one reason why theorists like Scarr (1992) and Plomin and Daniels (1987) can find little evidence of environmental effects on cognitive development.

There is much reason for assuming, then, that the environment of cognitive development is a richly structured milieu rather than consisting of positive or negative environmental 'factors' separately boosting or retarding a universal quantitative trait. In the previous chapter I suggested that high IQ may simply be an accident of immersion in middle-class cultural tools (aspects of literacy, numeracy, cultural knowledge, and so on), and the general rise in IQ in many countries, and what happens to IQs among migrant populations, are evidence for this (Flynn 1987; Ogbu

1986; Block 1995). All this suggests that we badly need to get away from the 'summation of forces' model, to one in which the environment is made up of socially structured devices and cultural tools, and in which development consists of the acquisition of such cultural tools.

Gene–environment interactions

6 The phenotypic value (e.g. IQ score) is then just the sum of these gene 'charges' and environmental 'charges' – i.e. according to the 'summation of forces' model, described in Chapter 3.
7 Thus, we have to pretend that there are no 'gene–gene' or 'gene–environment' (or, indeed, 'environment–environment') interactions – i.e. each gene 'charge' will make pretty much the same contribution to the trait whatever other genes are present, whatever the environmental 'charges' in which it operates, and vice versa.

Naturally, if all the genes influencing the development of a characteristic do not act in a simple additive sense, then the elementary genetic model breaks down, and the behaviour genetic method will almost certainly give erroneous results. This has been perhaps the biggest threat of all to that method. Kempthorne (1978: 17) described the idea that characteristics affected by genetic variation can be treated as a simple sum of the effects of single genes as 'a most remarkable defect'. Even in well-controlled animal studies the assumption has been problematic. As Barton and Turelli (1989: 342) put it in an extensive review, the 'additive gene action model is the basis for most quantitative genetic theory, and yet it is hard to know how often it applies'.

The behaviour genetic method as applied to humans by psychologists has to assume that there will be no such interactions because, if there are, this will completely ruin the calculus on which so much depends, including the interpretation of kinship correlations. We could no longer have 'expected' resemblances of 0.5 for DZ twins, parent–child pairs, and so on, because the expression of the different genes will be deviated to some extent, depending on the environment of development. They may, for example, converge so as to make an expected correlation of 1.0 not unreasonable, even for DZ twins; or interactions may work in the opposite direction to reduce the expected correlation. Moreover, the existence of gene–environment interactions would make the simple statistical partition of variance dubious because there will be an additional source of variance, the magnitude of which, as a proportion of the total variance, is entirely unknown: what usually happens is that it is simply lumped in with the genetic variance, thus possibly swelling the latter enormously, but misleadingly.

The gravest problem is that current methods for estimating heritability in human behaviour genetics *cannot even detect* gene–environment interactions. Apart from needing to know clearly in advance who has which genes for the trait in question, *and* the detailed effects of different environments, the test requires impossibly large samples. Testing for gene–environment interactions in humans 'is difficult statistically . . . and empirical attempts to find such interactions in the realm of behaviour have so far yielded little success' (Plomin 1994: 106; see also Wahlsten 1990).

This issue is repeatedly glossed over or dodged among the behaviour geneticists of IQ. Not surprisingly, Wachs (1992: 385) talks of 'decades of attempting to ignore the existence of interactions'. Grayson (1989) points out how the possibility of non-additive genetic interactions could render interpretations of conventional twin designs substantially biased. This is not a minor matter: on the contrary, it is absolutely central to nature–nurture arguments. Of course, the term 'interaction' has become something of a shibboleth in psychology so that nearly everyone happily subscribes to it. This is quite right, We saw in Chapter 2 that even in fruitflies, up to 90 per cent of genes may actually be regulatory genes, and 95 per cent of the DNA constituting genes may be regulatory, with the sole function of mediating such interactions. Animal work, in which truly controlled experiments can reveal them, suggests that gene–environment interactions are quite common even in traits with underlying genetic variability (e.g. Scott 1987; Kinsley and Svare 1987).

Behaviour geneticists repeatedly remind us of how genes and environments 'interact'. Plomin (1994), for example, several times reminds us of the 'interplay' between genes and between genes and environments but then quotes heritability estimates allegedly showing huge amounts of *additive* gene variance, as if unaware of the paradox involved. The APA group (Neisser *et al.* 1996) is also quick to remind us about gene–environment interactions, just in case anyone is still naive about them. *But then they go on to support data interpretation which assumes that those interactions do not exist.*

Such contradictions appear to justify Oyama's (1985: 5, 9) questions about the use of the term 'interaction': 'How does it manage to be virtually universally adopted and thus to lend itself to such radically different approaches?' She also expressed her amazement at 'the way people declare their rejection of the nature–nature opposition, sound the rallying cry of interactionism, then proceed to discuss genetically encoded instructions, and environmentally shaped traits'. The point (again, I'm afraid) is that we really cannot have it both ways. Either gene–environment interactions do not exist, or the classic correlation data fit a 'genetic model' that is quite untenable – which means that we should look for some alternative, more sensible model for twin and other kinship correlations. Either way, the assumption of non-interactions, most central to current behaviour genetics methods, appears also to be the one that is most obviously false.

Gene-environment covariance

There is yet another facet of the relationship between genes and environments which poses threats to the simple 'resemblance of relatives' calculus. Further sources of variance theoretically arise if 'favourable' genes tend to occur with favourable environments, and conversely for 'less favourable' genes (I put the terms in quotation marks because it should now be clear that this is a terrible way to talk about both genes and environments, especially for cognitive abilities). For example, it is considered plausible that children with genetically determined high IQ, say, will also enjoy an environment conducive to development of IQ because their parents will also tend to have genetically determined high IQ. They get a double dose of advantage. Such 'gene–environment correlation' will naturally spread the scores out further, increasing the variance (there is now a proportion of the total variance due to gene–environment correlation). This is a 'passive' covariance.

A more 'active' covariance is said to occur when a high IQ child, say, actually seeks out the kinds of stimuli, problem situations, or whatever, that 'help' the development of IQ (Scarr and McCartney, 1983). A 'reactive' covariance occurs when parents and teachers, say, respond to what they see as cues or symptoms of intelligence by giving the individual displaying them extra instruction, or channels them into certain activities. So again, within the terms of their own model, behaviour geneticists have a source of variance which is not estimated, but simply lumped in as part of the 'genetic' variance, possibly considerably distorting estimates of heritability.

Once more, such distortion cannot be checked because of the methodological hurdles mentioned above. As Plomin, DeFries, and Loehlin (1977: 321, quoted by Block 1995: 120) have conceded, 'It will probably never be possible to assess completely the effects of active and reactive genotype–environment correlations'. The reasons for this, as Block explains, are that

> reactive and active covariance cannot be measured without specific hypotheses about how the environment affects IQ. And it is just *a fact about IQ that little is known* about how the environment affects it. So reactive and active covariance is on the whole beyond the reach of the empirical methods of our era's 'behaviour geneticists', for those methods do not include an understanding of what IQ *is* . . . or how the environment affects it.
>
> (Block 1995: 118, Block's emphasis)

The genes → environments argument

Behaviour geneticists have described another supposed source of gene–environment covariance. Take, as an example, the study of Plomin, DeFries, and Fulker (1988) in which elderly reared-apart twins were asked for their perceptions of the environment in which they had been reared, based on the Family Environment Scales. This included eight factors such as 'cohesion', 'expressiveness', 'conflict', and 'control' as impressions of the family environment they had experienced as children. The finding that attracted most interest was a moderate positive correlation for both sets of twins on most of these factors, with that for MZs being higher than that for DZs. In other words, MZ twins reported greater similarity of rearing environment *even though members of pairs of both MZ and DZ twins had actually been brought up in different families.* This could, of course, be due to the cultural attribution of ability on the basis of appearance, mentioned hitherto. But Plomin (1994: 54), in keeping with the general tendency to interpret any MZ–DZ differences as 'genetic', indeed interprets these as 'genetic effects' in that 'members of the two families responded similarly to genetically influenced characteristics of the separated twins or that the twins contributed to similar family dynamics'. Scarr and McCartney (1983: 425) put it more strongly when they argue that 'genes drive experience'.

This new version of genetic determination, in which the 'genes' of children determine their environments, no matter what parents or others try to do, leads directly to all kinds of pessimistic practical implications. For instance, 'it is not easy to intervene deliberately in children's lives to change their development' and 'interventions have only temporary and limited effects' (Scarr 1992: 16). It also 'challenges parents' beliefs about their impact on their children's development': the 'more accurate' information now available (i.e from flawed twin and adoption studies) should 'alleviate needless sacrifices and emotional turmoil on their part'. In other words, if they fail in bringing up their children, parents should not feel guilty because the fault lies in the genes they have unwittingly bequeathed their children.

As noted in Chapter 1, there appears to be no end to the power of genes once they have been accorded such cognitive-like properties. Note the reductionist model, yet again, in these arguments – that the behaviour that does the evoking is merely a 'carrier' for regulations at the genetic level, and not the consequence of development at a totally different regulatory level (e.g. the socio-cognitive one). As put in a more recent paper (DiLalla and Bishop 1996: 540), we can expect 'to see mothers providing genes . . . that would elicit a certain behaviour in the children'. We do not behave according to our minds in their social situations: all our thoughts and behaviours are just shadowy reflections of what is going on in our genes as ultimate dictators.

No doubt there are other genetic interpretations yet to be squeezed out of kinship correlations. I suggest that a more objective stance might view them as socially extremely interesting phenomena, with, perhaps, important lessons in social psychology. This would require highly detailed and sensitive research programmes. But there has been little real interest in exploring possible explanations for those resemblances other than the genetic one which, as we have seen, are based on highly unlikely assumptions. The conviction of McGue *et al.* (1993) and others – that there can be no *possible* explanation for twin and adoptive child correlations other than a genetic one – is a little reminiscent of the 'obviousness' of the pre-Copernican universe.

The meaning and utility of heritability estimates

The 'genetic' interpretation of kinship correlations often becomes expressed numerically as a 'heritability' estimate, the proportion of total variance in a measured trait that is attributable to genetic variation. Heritability as a statistic arose in the context of previously unselected characters in plants and animals, thus showing much additive gene variance. The statistic served as an index of the likely fruitfulness of future breeding programmes. For example, if egg-laying in hens has a substantial heritability, then selecting, for breeding, only the most productive hens might help boost egg-production in all members of future generations.

In this context, it will be remembered, heritability is assessed with much control over both genes and environments, which is impossible with heritability estimation in humans. A particular estimate of heritability, is, of course, only applicable to the population and set of environments in which it has been assessed. Any gene–gene and gene–environment interactions, which we might expect in more evolved characters, would render unpredictable what would happen to the development of a character in another set of environments – and this includes any attempted interventions. Because of this a heritability is not applicable to individuals: it offers no explanation whatsoever for a given individual's cognitive ability.

Although presenting heritability estimates in the literature by the dozen, or in their hundreds, behaviour geneticists sometimes candidly acknowledge these restrictions. As Plomin (1994: 43–44) notes, 'Heritability . . . is a descriptive statistic that estimates . . . individual differences in a population, not behaviour of a single individual. . . . It describes "what is" rather than predicting "what could be" or "what should be"'.

Note that if a measure cannot predict 'what could be', then it cannot predict the outcome of any intervention, so it is scientifically useless. In the case of heritability estimates, then, we seriously need to ask why – even if they could be reliably obtained – psychologists attach so much significance to them. There may be several explanations for this. One purpose in clinging to heritability estimates is that they appear to furnish grounds for investigators to implicate 'genes', even though, as just seen, that implication appears to be a purely symbolic or idealistic one. It may be, as Block (1995: 121) suggests that 'part of the explanation is that behaviour genetics is a young field struggling for acceptance and funding, and heritability is a flag that attracts attention to it'.

Yet another explanation may be the common confusion that arises between causes of 'single-gene' defects in important characters and causes of population *variation* in a quantitative character. The APA group appears to fall into this trap when it suggests that 'data on heritabilities may give [psychologists] ideas about what to look for and where and when to look for it' (Neisser *et al.* 1996: 86). What they seem to be hinting at is where to look for pathological gene effects. We are all aware of deleterious single-gene effects, in which rare mutations in certain loci do not just exert minor effects proportional to their status of one allele among many, but a wholly *dis*proportionate, often devastating, effects on a character. Famous examples include phenylketonuria (PKU), galactosemia, and so on (see Wahlstrom 1990). Such a conspicuous, *discontinuous*, effect is precisely what we would expect from alteration of components in any highly integrated system. For instance, a 'mutation' in any component of your motor-car is likely to have a far more general effect than a small decrement in the car's performance, indicating (as if we didn't know already) that motor-cars are more than random collections of components each having a small positive or negative effect on performance.

Obviously, research on such defects, which may include studies of concordance in twins, is laudable work. But this is not *heritability* estimation in the sense of estimating additive genetic variation underlying a continuous quantitative character: reconstructing the cause of a (discontinuous) *defect* in individuals is logically quite different from estimating the additive genetic variance underlying the distribution of a *character* in a population. The latter tells us nothing about what to look for in individuals, nor where or when. That this categorical confusion is not unusual is seen in Plomin *et al.*'s (1994) suggestion that such single-gene defects may be just the 'tip of an iceberg' of general genetic variation in which alleles 'nudge' performance 'up or down'.

Given the practical inestimability and non-utility of heritability estimates in human psychology, it is difficult to work up a strong scientific case for attempting them at all. Nearly all the medical and other successful interventions that have been carried out with humans, from corrections of myopia to treatments for cancer, have been done without reference to heritabilities. All the genetic knowledge we have of the many different single-gene effects on cognitive traits (and, of course, more complex chromosomal abnormalities) affecting a tiny proportion of people – and whatever successful interventions have been devised for them – has been acquired without reference to heritability estimates. The danger is that the suggestions of the APA group will deflect research effort from identifying *real* causes of development and the *real* nature of characters like cognitive abilities to far more fruitless, if superficially exciting, targets.

Another possible explanation for the currency of heritability estimates, as already noted, may be the ease with which it is possible to slip from an isolated figure (the estimate) of very restrictive meaning into the more general conceptual furniture of genetic determination. Thus, almost all behaviour geneticists occasionally make the error of referring to heritability as 'the genetic determination of the characteristic being studied' (Rutter *et al.* 1990: 7). And while speaking the letter about heritability being irrelevant to intervention, Scarr (1992: 16) still insists, on the basis of heritability estimates for IQ, that 'it is not easy to intervene deliberately in children's lives to change their development'. This conceptual slide is, of course, not only illogical, but potentially dangerous.

IQ, 'race', and genetics

The cyclical process of attribution, from differences in appearance, self-presentation, and cultural affiliations, mentioned above, to genetic differences in 'intelligence', has been bad enough in the context of children in general. But it has also led, throughout the history of behaviour genetics, to a far nastier thesis, that of scientific racism. Almost all behaviour geneticists explicitly disavow such a thesis, but it is one that lurks immanently in the assumptions employed in the standard model of genetic determination of cognitive abilities. In consequence, scientific racism has repeatedly broken to the surface, usually, in more recent decades, with IQ scores and twin correlations as its main vehicles of expression.

As mentioned in Chapter 4, IQ testing arose in the USA and Britain precisely as the instrument of such ideology. Galton wanted such measures for his eugenic breeding programmes, to serve, as he put it, 'for the indications of superior strains or races, and in so favouring them that their progeny shall outnumber and gradually replace that of the old one' (1883). The IQ test very quickly became accepted as a measure of 'innate' cognitive ability, and thus of the genetic value of different groups of people. While in Britain, the issue became one of social class (the 'racial' superiority of the British having already been 'settled'), in America, with its large immigrant influx, the opposite was the case (Rose, Kamin, and Lewontin 1984). Thanks to the propaganda of test constructors like Goddard and Terman, sterilisation became common and, in 1924, immigration restrictions were placed on Slavs, Jews, Italians, and others on the basis of their supposedly innate low intelligence. As Rose, Kamin, and Lewontin (1984: 27) remind us, 'Ten years later the same argument was the basis for the German racial and eugenic laws that began with the sterilisation of the mentally and morally undesirable and ended in Auschwitz'.

After the Holocaust, scientific racism appeared to be in terminal disrepute. Yet in 1969, armed with the behaviour-genetics model of genetic determination, and Cyril Burt's (as it subsequently turned out, highly dubious) twin correlations, Arthur Jensen, an American psychologist, came to argue that the well-known mean IQ difference between American whites and blacks was, at least partly, due to genetic differences. The suggestion, and the support it has continued to attract in some quarters, all based on heritability estimations for IQ, led to a number of refutations both immediate (e.g. the contributions in Richardson and Spears 1973) and more recent (e.g. Block, 1995). The most recent expression of the same thesis is that of Herrnstein and Murray (1994) in a book entitled *The Bell Curve*, in deference to the 'natural' (but in reality wholly concocted) distribution of IQ.

All of the objections we have made about the incalculability and the futility of heritability estimates in human cognitive ability apply to this thesis. But there are at least two, even more damaging, objections. The first is the technical one that, even if it were possible to arrive at a reasonable heritability estimate for a group of people, it could not be applied *across* groups, in which different (but unknown) developmental conditions may prevail. The second is simply that human 'races' do not exist in any biologically meaningful sense: rather, the 'races' we talk about, and the distinguishing features we focus on, are *socially* defined. These do not map out genetic differences between groups in any *general* sense. Human groups, having evolved, seemingly, from a rather constrained stock and subjected to intense evolutionary pressures, are remarkably genetically alike. In addition, humans have been spectacular among species for their rates of migration and their inter-group breeding.

One result of this was shown by Lewontin (1974), who reported molecular genetic analyses with respect to the proteins produced by around 150 genes, using blood samples from large numbers of people from all around the world. The vast majority of genes (around 75 per cent) turn out to be identical in humans across the globe. The rest are polymorphic, in that different versions (alleles) of the gene exist with different frequencies in different groups (see Chapter 2). It is known from other studies that many or most of these variants will, themselves, be functionally indistinguishable. Remarkably, though, the distribution of none of them is such as to distinguish one human group from another. So 'inter-mixed' are they that any two people taken at random from anywhere in the world will exhibit at least 85 per cent of all the genetic variation existing in humans.

And naturally there is no information whatsoever that this genetic variation is associated with variation in cognitive ability.

'Discovery of a gene for IQ'

As we have seen in Chapter 3, the fact that complex traits involve many genes means that their separate effects are not 'visible' in the way that single-gene traits are. This alone has posed a formidable problem to those who want to resolve genetic variation underlying complex traits into simple effects of identifiable, separate genes. That problem has, of course, become several orders greater knowing what we now know about regulatory genes and the nature of the genome as a rich interactive system.

Historically, attempts to describe genetic variation underlying cognitive ability in humans have had to follow the 'biometrical' approach, in which the genotype is not 'measured' at all, but is only inferred from variation in the phenotype among family relatives. Statistical methods for the partitioning of variance into genetic, environmental, and other sources of variation were quite successful in plants and animals where properly controlled breeding experiments could be conducted. But such conditions are not applicable in the human situation. It is attempts to overcome this stricture, by using assumptions and approximations, in addition to misconceptions about the nature of the trait in question (i.e. cognitive ability), that have caused continual controversy, as we have just seen.

In recent years, though, the advent of molecular genetics promises to overcome at least some of the strictures of the biometric approach in a new research strategy. This strategy is said to 'yield direct detailed information about the genetic architecture of a quantitative trait: the number of loci involved, the frequencies and effects of their alleles and the types of loci (i.e. structural genes or regulatory genes)' (Boerwinkle, Chakraborty, and Sing 1986: 181). Although there are important conditions on the use of the approach (of which more below – see Boerwinkle *et al.* 1986: 181) it is, on the face of it, a dream come true for behaviour geneticists trying to establish the 'genetic basis' of differences in cognitive ability and who believe that cognitive ability can be treated as a simple quantitative character.

McClearn (1993: 47), for example, speaks of 'grand opportunities for the merging of molecular and quantitative genetics theories and methods' and suggests that the way is about to be opened 'for the application of the whole genetic armamentarium in the elucidation of mechanisms in polygenic systems'. Plomin *et al.* (1994: 116) put it even more starkly: such methods 'will make it possible to measure genotypes for individuals rather than relying on . . . inferences . . . based on familial resemblance', and that 'it seems clear that the field of behaviour genetics is at the dawn of a new era when molecular genetic techniques will be used to identify specific genes that contribute to the ubiquitous genetic influence found for behavioural dimensions as well as behavioral disorders'.

What is the basis of this excitement and how justified is it? First, it is necessary to warn the reader that, as applied to cognitive ability, the story entails the myriad assumptions about IQ, genes, environments, interactions, heritability, and so on, scrutinised already. But let us continue with the story as it has emerged.

In the 1920s it was pointed out how different values of a quantitative trait, although involving polygenes, seemed to be consistently associated with the *discrete* values (i.e. the different alleles) of a monogenetic trait, across generations. For example, different seed size (a quantitative trait) in beans was consistently associated with distinct categories of seed coat pigmentation (a discrete monogenetic character; Tanksley 1993). Thus, it seemed likely that the monogene was 'linked' to at least one of the polygenes (i.e. situated close to it on the same chromosome) so that they segregated together, only rarely being split by recombination in the formation of gametes. Since the identity of the monogene is revealed directly in its phenotype, whereas that is not the case for any of the polygenes, it was suggested that, by segregating together, the genes of one could be used as a 'marker' for the genes of the other. This, then, suggested the possibility of using successions of such 'marker genes' to map the loci underlying a quantitative trait (so-called Quantitative Trait Loci or QTL). This would, at least in theory, seem to solve the problem of genes for a quantitative character not being 'visible' in the way that single-gene, Mendelian traits are.

It was soon realised, not surprisingly, that the chances of finding a sufficient number of such morphologically distinct marker genes were extremely remote (Tanksley 1993). But ways of overcoming this constraint arose in the 1970s. To understand this you have to remember from Chapter 1 how genes are arranged on pairs of chromosomes, each occupying a specific locus, and sometimes forming structurally different versions called alleles. You will also remember how the existence of different alleles can be shown, even where there is no phenotypic variation (as is usually the case) because their products, called allozymes or isozymes, carry slightly different electric charges and can be separated on electrophoretic gels. This offered the prospect of many more markers being available for QTLs. And this availability has recently been augmented by direct chemical analysis of underlying DNA, which shows even more allelic variation.

These methods have solved the problem of limited marker abundance, so that today linkage maps covering a number of quantitative traits have been established in experimental animals, although the expression 'Quantitative Trait Loci' is used to reflect the fact that they might include groups of unknown size rather than single polygenes.

As might be expected, for many traits with many loci, assaying suitably sized populations for marker genes becomes time-consuming and expensive. An alternative approach is to measure all individuals in the population for the trait and then use only those at the extreme ends of the distribution. If the allelic frequencies at a molecular marker locus differ significantly between these two groups it is usually inferred that a marker has been identified (Tanksley 1993).

The appeal of this method for the human behaviour geneticist who sees IQ as a simple quantitative trait, determined by the sum of polygenes, is obvious. Plomin (1994: 164) claims that, 'The breathtaking advances in molecular genetics during the past two decades have made it possible to begin to identify some of the many genes likely to be responsible for genetic variance in complex traits'. Although there are numerous problems in implementing the approach with complex traits in humans (see Tanksley 1993) it is this method of 'allelic association' that Plomin *et al.* (1994) have used, on IQ, with results reported in the journal *Behavior Genetics*.

Prefigured in national newspapers as the impending discovery of the first 'gene for IQ', the editors of the journal were obviously deeply impressed. Getting a paper considered by referees, and the editorial board, can take up to six months in most learned journals. Then corrections are usually necessary, and, if acceptable, the paper has to take its place in the queue for final printing and publication. Authors can usually be considered fortunate if the paper is published a year after first submission. The paper of Plomin *et al.* was submitted on 30 September 1993, corrected by 20 October and published in the second issue of 1994 – a degree of haste, attention, and prioritisation that is reminiscent of the paper by Crick and Watson in 1955 on the discovery of the structure of DNA!

Plomin *et al.* took twenty-four 'high IQ' subjects and eighteen 'low IQ' subjects (all six to twelve years old), took blood from them, and sent the samples to biochemical labs for analysis of the frequencies of alleles of DNA markers at sixty loci. The hope was, of course, that one group would have more of some markers than the other group, and these might actually identify alleles that produce the differences in IQ. The selection of these markers was largely based on guesswork, though they are said to be 'in or near genes likely to be relevant to neural functioning'. Likewise, it is not known whether the allelic variants have any functional significance, although, as mentioned above, this is usually not the case. A 'replication' sample (ages not given) of twenty-seven 'high *g*' subjects and seventeen 'low *g*' subjects was also studied.

What was found? Well, not very much at all. Significant associations between allele frequencies and IQ grouping were few and not replicated: 'none of the the five two-allele markers that yielded significant associations in the original sample also reached significance in the replication sample' (Plomin *et al.* 1994: 115). (Note that, in this number of comparisons, two statistically significant associations would have been expected on the basis of chance alone.) 'Similarly for the multiple allele markers, 3 of 26 comparisons were significant in the original sample. Again, none also reached significance in the replication sample' (Plomin *et al.* 1994: 116). Quite understandably, the authors declare that 'we will not conclude that we have identified a significant QTL for *g* until we have cleanly replicated an association in an independent sample' (Plomin *et al.* 1994: 116). (In an earlier publication, vaguely heralding significant findings, Plomin and Thompson (1993: 77) said that, 'Although interesting preliminary results are beginning to emerge, we have agreed not to publish these results until we have replicated them in an independent sample'.)

The genic imperative

No doubt this 'new dawn' for behaviour geneticists will shine for some time to come. (In a more recent study, for example, similar claims have been made about genetic markers for memory, even though psychologists have as much difficulty describing memory as they do IQ, that nothing is known about the functional significance for memory of the alleles in question, that no *causal* connection is indicated, and the associations were few, weak, and very confusing (Nilsson *et al.* 1996).) My worry is that this is how scientific myths are created. However unintentionally, the impression created for psychologists, journalists, and members of the general public is that 'real' genes have been discovered as the causes of individual differences in a 'real' quantitative trait, when, of course, no such thing has happened. As we saw in Chapter 4, far from being a clearly identifiable 'behavioural dimension', psychologists cannot agree about what IQ *is*; and any resemblance it has to a quantitative character is one created in the technology of test construction.

I want to stress that this pessimism about the utility of the approach with IQ (a non-quantitative trait, with totally unknown underlying genetic variation) is not a dismissal of the approach generally. This is now well proven with genuine quantitative traits (Tanksley 1993; Edwards, Stuber, and Wendell 1987). But we do need to remember that human cognitive abilities are part of the most evolved system of regulations yet appearing. Genes for evolved systems do not exist as independent entities, but as a flexibly regulated 'team', in which the identity of any one is meaningless outside the context of the whole. It is surely self-contradictory for anyone aware of such organisation to claim, as do Plomin *et al.,* that

> rare alleles that drastically disrupt cognitive development are likely to be just the most easily noticed tip of the iceberg of genetic variability. It seems reasonable to expect that many more

> alleles nudge development up as well as down . . . and do not show such striking effects.
>
> (Plomin *et al.* 1994: 108)

This puts into perspective what the implications would be if we *did* discover a significant, replicable, allelic association for IQ, and even *if* IQ could be considered a simple quantitative trait. For a start, such a correlation could not be interpreted as causal until such a causal connection had been demonstrated *experimentally*. How such studies are to be carried out is not explained. But the situation is much more complicated than that. As Scarr and Carter-Saltzman (1983: 815), themselves prominent contributors to the behaviour-geneticist programme, put it: 'Even if the gene action pathways for the hundreds of loci were known, and the systematic interactions among the loci known, the relationship between genotype and phenotype would still be rendered indeterminate for individuals by idiosyncratic and environmental events'.

Many other authors have criticised such a starkly atomistic view of the gene.

> Genes in populations do not exist in random combinations with other genes. . . . The fitness of a single locus ripped from its interactive context is about as relevant to real problems of evolutionary genetics as the study of the psychology of individuals isolated from their social context is to an understanding of (human) sociopolitical evolution. In both cases context and interaction are not simply second order effects to be superimposed on a primarily monadic analysis. Context and interaction are of the essence.
>
> (Lewontin 1974: 318)

Ironically, such scepticism is fulsomely reinforced by molecular genetic studies. 'As we scrutinize the molecular landscape with increasing sophistication, rather than discovering a lower level of fundamental control, regulation appears instead to recede rapidly into higher levels of organization' in which 'what is being uncovered at the molecular level is not some fundamental level of invariant, ultimate control, but complex interactive systems with cascading levels of organization and numerous feedback systems' (Rollo 1995: 80, 6).

These views are not encouraging about the search for single alleles, acting independently and additively to 'nudge' IQ (or memory, or whatever) up or down. This is not a put-down of genetic research on cognitive abilities. What worries me is the perpetuation of the idealism about genes (and about the nature of cognitive ability and the nature of environments relevant to it) that I complained about in Chapter 1. If we had an agreed and clear understanding of the nature of cognitive ability, it would be intrinsically interesting to have a clear picture of the role of genes (along with everything else) in its development. The problem is that we do not currently have such understanding, and that, surely, should be our number one priority.

For these reasons, any successful genetic research in cognitive abilities is likely to be confined to the elucidation of rare deleterious alleles, affecting relatively few people, whose population distribution is discontinuous, not the tail-end of a spread of 'positive' and 'negative' alleles in the population generally (Wahlstrom 1990; Rutter, Simonoff, and Silberg 1993). For such enquiries twin studies could be conceivably useful. Indeed, a number of relatively rare cognitive defects could, after suitably meticulous characterisation, fall into that class. On the other hand, we need to stress the strictures already mentioned. We need to be clear about the nature of the character and the nature of variation in it. We need to be clear about the *nature* of any defect (which is much clearer in some cases than in others) and appreciate how examining the genetics of a defect is logically different from establishing the origins of its natural variation. And we need to remember that a correlation is not a cause. Looking for idealistic 'genes' through hazy correlational analyses, as a way of bypassing clear characterisation of a trait and defects in it, will, in my view, only lead to further confusion.

In short, if the further discovery of rare single-gene effects can lead to the amelioration of obvious (and obviously rare) cognitive defects, all well and good. The more general problem of promotion of cognitive abilities, though, lies, in my opinion, at an evolutionarily quite different level. I shall return to this matter in the next chapter.

Ethics and common sense

In the final comments of their 'DNA markers for IQ' paper, Plomin *et al.* (1994: 117) worry about the ethical implications of their work. 'As is the case with most important advances, it will raise new ethical issues'. And as Scarr (1993: 1350) concurs, 'it is too late for social scientists to prevent advances in genetic knowledge. Perhaps becoming more educated in the potentially positive uses of such information will allay some fears'. Plomin *et al.* quote Muller-Hill (1993: 492): 'the isolation of the first gene involved in determining "intelligence" (whatever that is) will be a turning point in human history'. So, as the note in parentheses confesses, we will have isolated something mysteriously associated with something, the nature of which we are totally ignorant of: this seems a strange turning point!

Colleagues of the British members of the Plomin *et al.* (1994) research team have expressed a similar worry. The paper 'has resulted in considerable debate in our Department of Medical Genetics in Cardiff', a representative of the Department writes in a letter to the editor of *Behaviour Genetics* (Harper 1995: 197). After expressing concern that the ethical issues should receive the widest possible airing, he adds that

> The practical implications and future implications of this research are uncertain. The authors point out clearly that their work is unlikely to form the basis of future tests of prediction because of the probable small effect of individual loci. . . . However . . . there could well be pressure to apply (or misapply) any marker . . . even without an adequate scientific basis for doing so. . . . Given the tendency to sensationalise and trivialise . . . there is a distinct possibility . . . of adverse publicity . . . that might increase the fears of the public about the work of scientists and clinicians in this field, especially given the history of eugenic abuse in psychiatric genetics. . . . As a result of the discussions and of the continuing concerns within the Department of Medical Genetics, I have decided, in consultation with colleagues, that this department cannot be involved in or associated with the work until the ethical issues are clarified and hopefully resolved.
>
> (Harper 1995: 198)

The real pity about this letter is that there would not *be* an ethical dilemma if geneticists involved in this enterprise were clear about the nature of the tools (IQ), the murkiness of the concepts of ability, and the assumption-laden genetic models, which psychologists had induced them into sharing. If there is an ethical dilemma at all it is to stop raising spectres of knowledge that we do not have, but which, in the process (and no doubt unconsciously) help perpetuate a 'culture of pessimism' about vast numbers of the world's children. Let me remind the reader that not one *positive* fruit of the pessimistic behaviour-genetic culture surrounding cognitive ability has yet been achieved after most of a century of trying – though, as I will argue again in Chapter 6, there have been many *negative* consequences. I believe this is the case because the assumptions on which that whole culture has been based are seriously awry. So, perhaps, the time is long overdue for applying a far more *optimistic* culture to the promotion of cognitive abilities. In the final chapter, I suggest that the psychological bases of such optimism already exist.

References

Bandura, A., Barbaranelli, C., Caprara, G.V., and Pastorelli, C. (1996). Multifaceted impact of self-efficacy beliefs on academic functioning. *Child Development*, 67, 1206–1222.

Barton, N.H. and Turelli, M. (1989). Evolutionary quantitative genetics: how little do we know? *Annual Review of Genetics*, 23, 337–370.

Baumrind, D. (1993). The average expectable environment is not good enough: a response to Scarr. *Child Development*, 64, 1299–1317.

Beckwith, J., Geller, L., and Sarkar, S. (1991). Letter. *Science*, 253, 191.

Berschied, E. and Walster, E. (1978). *Interpersonal Attraction.* Reading, MA: Addison-Wesley.

Block, N. (1995). How heritability misleads about race. *Cognition*, 56, 99–128.

Boerwinkle, E., Chakraborty, R., and Sing, C.F. (1986). The use of measured genotype information in the analysis of quantitative phenotypes in man. *Annals of Human Genetics*, 50, 18–194.

Bouchard, T.J. Jr (1983). Do environmental similarities explain the similarity in intelligence of identical twins reared apart? *Intelligence*, 7, 175–184.

Bouchard, T.J. Jr, Lykken, D.T., McGue, M., Segal, N.L., and Tellegen, A. (1990). Sources of human psychological differences: the Minnesota study of twins reared apart. *Science*, 250, 223–250.

Bouchard, T.J. Jr and McGue, M. (1981). Familial studies of intelligence: a review. *Science*, 212, 1055–1059.

Bouchard, T.J. Jr and Propping, P. (1993). Twins: nature's twice-told tale. In T.J. Bouchard Jr and P. Propping (eds) *Twins as a Tool of Behaviour Genetics.* Chichester: Wiley.

Bradley R.H. (1994). The HOME inventory: review and reflections. In H.W. Reese (ed.) *Advances in Child Development and Behavior*, Vol. 25. San Diego: Academic Press.

Brodzinsky, D.M., Singer, L.M., and Braff, A.M. (1984). Children's understanding of adoption. *Child Development*, 55, 869–878.

Bronfenbrenner, U. (1986). Ecology of the family as a context for human development. *Developmental Psychology*, 22, 723–742.

Bruner, J.S. (1974). *Beyond the Information Given: Studies in the Psychology of Knowing*. London: Allen & Unwin.

Burt, C. (1959). Class differences in intelligence. *British Journal of Statistical Psychology*, 12, 15–33.

Caldwell, B.M. and Bradley R.H. (1984). *HOME Observation for Measurement of the Environment.* Little Rock: University of Arkansas at Little Rock.

Cicero (1975). *Volume XXI – De Officius* (trans. W. Miller). London: Heinemann.

Corder, E.H., Saunders, A.M., Strittmatter, W.J., Schmechel, D.E., Gaskell, P.C., Small, G.W., Roses, A.D., Haines, J.L., and Pericak-Vance, M.A. (1993). Gene dose of apolipoprotein E type 4 allele and the risk of Alzheimer's disease in late onset families. *Science*, 261, 921–923.

Detterman, D.K. and Sternberg, R.J. (1986). *What is Intelligence? Contemporary Viewpoints on its Nature and Definition.* Norwood, NJ: Ablex.

DiLalla, L.F. and Bishop, E.G. (1996). Differential maternal treatment of infant twins: effects on infant behaviors. *Behavior Genetics*, 26, 535–542.

Eaves, L.J., Eysenck, H.J., and Martin, N.G. (1986). *Genes, Culture and Personality.* London: Academic Press.

Edwards, M.D., Stuber, C.W., and Wendell, J.F. (1987). Molecular marker facilitated investigations of quantitative trait loci in maize – I: numbers, genomic distribution and types of gene action. *Genetics,* 116, 113–125.

Everson, M.D., Sarnat, L., and Amron, S.R. (1981). Day care and early socialisation: the role of maternal attitude. In R.C. Ainslie (ed.) *Quality Variations in Day Care*. New York: Praeger.

Flavell, J.H. (1992). Cognitive development: past, present and future. *Developmental Psychology*, 28, 998–1005.

Flynn, J.R. (1987). Massive IQ gains in fourteen nations: what IQ tests really measure. *Psychological Bulletin*, 101, 171–191.

Galton, F. (1883). *Inquiry into Human Faculty and its Development*. London: Macmillan.

Gillis, J.S. (1982). *Too Tall, Too Small*. Champaign, IL: Insitute for Personality and Ability Testing.

Goodman, R. and Stevenson, J. (1989). A twin study of hyperactivity – II: the aetiological role of genes, family relationships and perinatal adversity. *Journal of Child Psychology and Psychiatry*, 30, 691–709.

Grayson, D.A. (1989). Twins reared together: minimizing shared environmental effects. *Behavior Genetics*, 19, 593–604.

Harper, P.S. (1995). DNA markers associated with high versus low IQ: ethical considerations (Letter to the Editor). *Behavior Genetics*, 25, 197–198.

Hajal, F. and Rosenberg, E.G. (1991). The family life cycle in adopted families. *American Journal of Orthopsychiatry*, 61, 78–85.

Hearnshaw, L. (1979). *Sir Cyril Burt: Psychologist*. Ithaca, NY: Cornell University Press.

Herrnstein, R.J. and Murray, C. (1994). *The Bell Curve*. New York: Free Press.

Honzik, M. (1957). Developmental studies of parent–child resemblance in intelligence. *Child Development*, 28, 215–228.

Horn, J.M., Loehlin, J.C., and Willerman, L. (1979). Intellectual resemblances among adoptive and biological relatives: the Texan adoption study. *Behavior Genetics*, 19, 177–207.

Horn, J.M., Loehlin, J.C., and Willerman, L. (1986). The TAP ten years later. Symposium at the Sixteenth Annual Meeting of the Behavior Genetics Association, 18 June, Honolulu, Hawaii.

Jackson, J.F. (1993). Human behavioral genetics, Scarr's theory, and her views on interventions: a critical review and commentary on their implications for African American children. *Child Development*, 64, 1318–1332.

Kamin, L. (1974). *The Science and Politics of IQ*. New York and London: Wiley.

Kamin, L. (1981). Commentary. In S. Scarr (ed.) *Race, Social Class, and Individual Differences in IQ*. Hillsdale, NJ: Erlbaum.

Kempthorne, O. (1978). Logical, epistemological and statistical aspects of nature–nurture data interpretation. *Biometrics*, 34, 1–23.

Kimble, G.A. (1993). Evolution of the nature–nurture issues in the history of psychology. In R. Plomin and G.E. McClearn (eds) *Nature, Nurture and Psychology*. Washington, DC: American Psychological Association.

Kinsley, C. and Svare, B. (1987). Genotype modulates prenatal stress effects on aggression in male and female mice. *Behavioral and Neural Biology*, 47, 138–150.

Lamb, M.E, (1987). Niche picking by siblings and scientists. *Behavioral and Brain Sciences*, 10, 30.

Langlois, J.H. (1986). From the eye of the beholder to behavioral reality: Development of social behaviors and social relations as a function of physical attractiveness. In C.P. Herman, M. Zanna, and E.T. Higgins (eds) *Physical Appearances, Stigma and Social Behaviour.* Hillsdale, NJ: Erlbaum.

Lerner, L.M. (1968). *Heredity, Evolution and Society.* San Francisco: W.H. Freeman.

Lewontin, R.C. (1974). *The Genetic Basis of Evolutionary Change.* New York: Columbia University Press.

Lloyd, B. and Duveen, G. (1991). The reconstruction of social knowledge in the transition from the sensorimotor to conceptual activity: the gender system. In M. Woodhead, R. Carr, and P. Light (eds) *Becoming a Person.* London: Routledge in association with The Open University.

Locurto, C. (1991). *Sense and Nonsense about IQ.* New York: Praeger.

Loehlin, J.C. and Nichols, R.C. (1976). *Heredity, Environment and Personality.* Austin: University of Texas Press.

McClearn, G.E. (1993). Behavioral genetics: the last century and the next. In R. Plomin and G.E. McClearn (eds) *Nature, Nurture and Psychology.* Washington, DC: American Psychological Association.

McGue, M., Bouchard, T.J., Iacona, W.G., and Lykken, D.T. (1993). Behavioral genetics of cognitive ability: a life-span perspective. In R. Plomin and G.E. McClearn (eds) *Nature, Nurture and Psychology.* Washington, DC: American Psychological Association.

Mackintosh, N.J. (1995). *Cyril Burt: Fraud or Framed?* Oxford: Oxford University Press.

Mascie-Taylor, C.G.N. (1993). How do social, biological and genetic factors contribute to individual differences in cognitive abilities? In T.J. Bouchard Jr and P. Propping (eds) *Twins as a Tool of Behaviour Genetics.* Chichester: Wiley.

Matheny, A.P., Wilson, R.S., and Dolan, A.B. (1976). Relations between twins' similarity of appearance and behavioral similarity: testing an assumption. *Behavior Genetics*, 6, 343–351.

Mayr, E. (1988). *Towards a New Philosophy of Biology: Observations of an Evolutionist.* Cambridge, MA: Harvard University Press.

Micceri, T. (1989). The unicorn, the normal curve and other improbable creatures. *Psychological Bulletin*, 25, 156–166.

Muller-Hill, B. (1993). The shadow of genetic injustice. *Nature*, 362, 491–492.

Neisser, U., Boodoo, G., Bouchard, T.J. Jr., Boykin, A.W., Brody, N., Ceci, S.J., Halpern, D.F., Loehlin, J.C., Perloff, R., Sternberg, R.J. and Urbina, S. (1996). Intelligence: knowns and unknowns. *American Psychologist*, 51, 77–101.

Nilsson, L.-G., Sikstrom, C., Adolfsson, R., Erngrund, K. Nylander, P.-O., and Beckman, L. (1996). Genetic markers associated with high versus low performance on episodic memory tasks. *Behavior Genetics*, 26, 555–562.

O'Brien, M., Johnson, J.M., and Anderson-Goetz, D. (1989). Evaluating quality in mother–infant interaction: situational effects. *Infant Behavior and Development*, 12, 451–464.

Ogbu, J. (1986). The consequences of the American caste sytem. In U. Neisser (ed.) *The School Achievement of Minority Children: New Perspectives.* Hillsdale, NJ: Erlbaum.

Oyama, S. (1985). *The Ontogeny of Information.* Cambridge: Cambridge University Press.

Pedersen, N.L., Plomin, R., Nesselroade, J.R., and McClearn, G.E. (1992). A quantitative genetic analysis of cognitive abilities during the second half of the life span. *Psychological Science*, 3, 346–353.

Plomin, R. (1994). *Genetics and Experience*. London: Sage.

Plomin, R. and Daniels, D. (1987). Why are children in the same family so different from one another? *Behavioral and Brain Sciences*, 10, 1–16.

Plomin, R. and DeFries, J.C. (1983). The Colorado Adoption Project. *Child Development*, 54, 276–289.

Plomin, R., DeFries, J.C., and Fulker, D. (1988). *Nature and Nurture in Infancy and Early Childhood.* New York: Cambridge University Press.

Plomin, R., DeFries, J.C., and Loehlin, J.C. (1977). Genotype–environment interaction and correlation in the analysis of human behaviour. *Psychological Bulletin*, 84, 309–322.

Plomin, R. and Loehlin, J.C. (1989). Direct and indirect IQ heritability estimates: a puzzle. *Behavior Genetics*, 19, 331–342.

Plomin, R., McClearn, G.E., Smith, D.L., Vignetti, S., Chorney, M.J., Chorney, K., Venditti, C.P., Kasarda, S., Thompson, L.A., Detterman, D.K., Daniels, J., Owen, M., and McGuffin, P. (1994). DNA markers associated with high versus low IQ: the IQ Quantitative Trait Loci (QTL) project. *Behavior Genetics*, 24, 107–118.

Plomin, R. and Thompson, L.A. (1993). Genetics and high cognitive ability. In *CIBA Foundation Symposium 178: The Origins and Development of High Ability*. Chichester: Wiley.

Richardson, K. and Bynner, J.M. (1984). Intelligence: past and future. *International Journal of Psychology*, 19, 499–526.

Richardson, K. and Spears, D. (1973). *Race, Culture and Intelligence.* Harmondsworth: Penguin Education.

Roff, D.A. and Mousseau, T.A. (1987). Quantitative genetics and fitness. *Heredity*, 58, 103–118.

Rollo, D.C. (1995). *Phenotypes: Their Epigenetics, Ecology and Evolution.* London: Chapman and Hall.

Rose, S., Kamin, L.J. and Lewontin, R.C. (1984). *Not in Our Genes.* Harmondsworth, Penguin.

Rose, R.J. and Kaprio, J. (1988). Twin similarity and frequency of contact. *Behavior Genetics*, 18, 308–309.

Rosenberg, E.G. and Horner, T.M. (1991). Birthparent romances and identity formation in adopted children, *American Journal of Orthopsychiatry*, 61, 70–77.

Rutter, M., Bolton, P., Harrington, R., Le Couter, A., Macdonald, H., and Simonoff, E. (1990). Genetic factors in child psychiatric disorders – I: a review of research strategies. *Journal of Child Psychology and Psychiatry*, 31, 3–37.

Rutter, M., Simonoff, E., and Silberg, J. (1993). How informative are twin studies of child psychopathology? In T.J. Bouchard Jr and P. Propping (eds) *Twins as a Tool of Behaviour Genetics.* Chichester: Wiley.

Schiff, M. and Lewontin, R. (1986). *Education and Class: The Irrelevance of IQ Genetic Studies*. Oxford: Clarendon.

Scarr, S. (1989). Protecting general intelligence: constructs and consequences for intervention. In R.L. Linn (ed.) *Intelligence: Measurement, Theory and Public Policy.* Urbana: University of Illinois Press.

Scarr, S. (1992). Developmental theories for the 1990s: development and individual differences. *Child Development*, 63, 1–19.
Scarr, S. (1993). Biological and cultural diversity: the legacy of Darwin for development. *Child Development*, 64, 1333–1353.
Scarr, S. and Carter-Saltzman, L. (1979). Twin method: defense of a critical assumption. *Behavior Genetics*, 9, 527–541.
Scarr, S. and Carter-Saltzman, L. (1982). Genetics and intelligence. In R.J. Sternberg (ed.) *Handbook of Human Intelligence.* Cambridge: Cambridge University Press.
Scarr, S. and McCartney, K. (1983). How people make their own environments: a theory of genotype → environment effects. *Child Development*, 54, 424–435.
Scott, J.P. (1987). Why does human twin research not produce results consistent with those from nonhuman animals? *Behavioral and Brain Sciences*, 10, 39–40.
Shields, J. (1962). *Monozygotic twins brought up apart and brought up together.* London: Oxford University Press.
Snyderman, M. and Rothman, S. (1990). Survey of expert opinion on intelligence and aptitude testing. *American Psychologist*, 42, 137–144.
Tanksley, S.D. (1993). Mapping polygenes. *Annual Review of Genetics*, 27, 205–233.
Taylor, H.F. (1980). *The IQ Game.* New Brunswick, NJ: Rutgers University Press.
Terwogt, M.M., Hoeksma, J.B., and Koops, W. (1993). Common beliefs about the heredity of human characteristics. *British Journal of Psychology*, 84, 499–503.
Trevarthen, C. and Logotheti, K. (1989). Child and culture: genesis of cooperative knowing. In A. Gellatly, D. Rogers, and J. A. Sloboda (eds) *Cognition and Social Worlds.* Oxford: Clarendon.
Wachs, T.D. (1983). The use and abuse of environment in behavior genetic research. *Child Development*, 54, 396–407.
Wachs, T.D. (1992). *The Nature of Nurture.* London: Sage.
Wagner, R.K. (1994). Context counts: the case of cognitive ability testing for job selection. In R.J. Sternberg and R.K. Wagner (eds) *Mind in Context: Interactionist Perspectives on Human Intelligence*. Cambridge: Cambridge University Press.
Wahlsten, D. (1990). Insensitivity of the analysis of variance to heredity–environment interaction. *Behavioral and Brain Sciences*, 13, 109–120.
Wahlstrom, J. (1990). Gene map of mental retardation. *Journal of Mental Deficiency Research*, 34, 11–27.
Woodhead, M. (1996). *In Search of the Rainbow.* The Hague: Bernard van Leer Foundation.

6

POTENTIAL IN THE COGNITION-CULTURE COMPLEX

Introduction

Why are the myriad assumptions scrutinised throughout this book so widely embraced and accepted by both psychologists and the general public? Why do we believe that cognitive ability can be conceived like any quantitative character, as far as modelling causes of its development and individual differences in it are concerned? Why do we believe that it can be defined in terms of fixed inner structures or constraints that 'fit' some corresponding stable demands of the environment? Why do we believe that it is 'normally distributed' in the population, when most measurable traits are not? Why do we reduce cognitive ability to a simple 'two-layer' model in which what we 'see' on the 'outside' is just a smudged expression of the genes on the 'inside'? Why do we continue to believe in IQ when we recognise that nearly all children know far more and can think in far more complex ways than is ever required in an IQ test, and when a test score predicts nothing that is not actually built into the tests? Why do we believe that genes and environments can be analytically atomised into independent elements having additive effects, when even those who construct genetic models based on those assumptions stake their claims as vibrant 'interactionists'?

I believe that we tend to accept these assumptions – and the many others analysed in this book – because they form part of an integrated 'worldview' that is difficult to shake for many reasons. One reason is that they have an internal coherence sufficient to serve as a simple model that, like the pre-Copernican model of the sun and planets circling the earth, actually 'fits' everyday social experience. Such a model eases decision-making in our main institutions such as family, work, and school. In addition, attempting to construct and accommodate alternative models, especially one which flies in the face of common experience, is intellectually demanding, even when we know that *current* assumptions are invalid. Another reason may be that giving up those assumptions implies other consequences for society that may be awkward, possibly painful.

One way of questioning an inappropriate model or 'worldview' is to show that, in spite of a superficial sense of legitimacy, it is actually seriously dysfunctional for the vast majority of people, and for society as a whole. This is, of course, what we would expect from a model of human abilities based on shaky assumptions, both quasi-biological and psychological. Demonstrating such dysfunction in the worldview defined by the set of assumptions scrutinised so far is what I try to do in this final chapter.

Naturally, this cannot be an exhaustive review. However, my task is made easier because most of the assumptions in question are brought together in the deceptively simple concept of 'potential'. Indeed, the twentieth century, in all developed societies, has witnessed the setting up of a special institution which has that idea at the roots of its precepts and practices. This is the institution of education. So in the first part of this chapter I try to show how the worldview is dysfunctional by examining the 'working out' of human potential as commonly conceived in the education system.

It would be remiss of me to stop there, though. This book, so far, has been largely antithetical, and some alternative view is fairly demanded. One of the things I have really wanted to show is that, when parents, teachers, and others ask questions about how to fulfil our children's humanity by maximising their potentials, they are usually misconceiving the whole nature of human potential, and thus asking questions based on false assumptions. In the second part of this chapter, therefore, I try to elaborate on an alternative model of cognitive ability that is far more optimistic about human 'developmentability' in cognition. To do this I pick up on the 'levels of regulations' arguments put forward in Chapters 2 and 3, focus on socio-cognitive regulations, and, with illustrations, point again to the creativity of interactions when they are free from undue constraints.

Potential lost

We are all familiar with the image of the minority of humans, talented or 'gifted' by virtue of a fortunate allocation of genes, and effortlessly scaling heights of achievement throughout life by dint of their special potential. Promoted by the IQ test, and the theories of behaviour geneticists, such beings are almost as popular among psychologists as they are among members of the general public. Apart from seemingly confirming the accepted worldview of a natural rank of ability, these individuals are sometimes seen as the vehicles for all our hopes for the human species – and simultaneously as rationalisation for the more mundane efforts and achievements of the rest of us.

When the existence of these innately talented or gifted beings is scrutinised a little more closely, though, it turns out that they are probably as mythical as the unicorn. A large number of case studies and laboratory investigations has now been carried out on children and adults displaying exceptional ability in a wide variety of perceptual, cognitive, musical, artistic, and other domains. When they are compared with normal individuals it has proved impossible to find a residue of anything resembling 'innate talent' independent of experience. 'Efforts to specify and measure characteristics of talent that allow early identification and successful prediction of adult performance have failed' (Ericsson and Charness 1994: 744).

What *is* found is that such individuals have invariably been through periods of intense efforts to learn, involving high levels of commitment and self-sacrifice over long periods of time. In children it is also usually the case that parents have given up 'almost everything' (Feldman 1986: 122) to ensure that the specific ability is developed in their child or children. Instead of supporting the idea of mysterious congenital potential, such individuals are suggestive of just how far and how quickly general human potential can be manifested, given appropriate learning conditions. As Ericsson and Charness (1994: 744) put it in their review of all such studies, 'the evidence from systematic laboratory research . . . provides no evidence for giftedness or innate talents but shows that exceptional abilities are acquired often under optimal environmental conditions. . . . It is curious how little empirical evidence supports the talent view of expert and exceptional performance' (see also Howe 1990).

In spite of this difficulty, the image of the genetically bright and dull pervades our schools. Here, again, assumptions have been compressed into a single, widely used concept of 'potential'. Indeed, probably the most general and widespread idea in the education system is that of inherent *potential,* partly common to, and partly distinguishing, all children. One of the most earnest goals of education is frequently described as that of helping children attain their 'full potential', or their 'different potentials', so conjuring up the image of predetermined abilities simply waiting to be 'filled out' by proper educational and other experiences.

This view of innate potential mysteriously unfolding itself, with its biological undertones, still furnishes our most common metaphors for learning ability and individual differences in it. Thus, learning in the school is seen as a kind of Darwinian natural selection process in which those with 'strong' potential are sorted out from the 'weak' by a neutral curriculum. As a result, children's school achievements are seen, more than anything else, as the expression of their natural potential in making them particularly suited (again mysteriously) for the knowledge and skills they have acquired. In addition, tests and exams of many sorts are taken to be indicative of children's and young people's natural potentials, rather than merely current states of development, family and cultural background, interest and motivation. Concepts such as aptitudes, talents, and 'bright' and 'dull' children pervade the National Curriculum and discussions around it. And, of course, recent moves in Britain to establish and introduce tests that will identify eleven-year-olds with certain 'aptitudes', and thus channel them into privileged education, reflects a clinging to the same metaphor.

The perpetuation of essentially the same metaphor of potential across a century of state education explains why, in spite of 'great debate' after 'great debate', nothing has inherently changed within it. So fundamental is the metaphor of inherent potential that most people seem to be unable to entertain the *possibility* of it not being a valid one: rarely is this actually checked. This taken-for-grantedness must, again, reflect the durability of the assumptions underlying the whole notion of inherent individual potential.

When the validity of the idea of potential *is* checked, though, empirical support for it is difficult to find. For example, if doing well in education and afterwards *is* the ineluctable expression of an inherent potential, we would expect it to 'show' continuously. Yet research has repeatedly indicated that performances that are taken as a signal of potential are, at best, only very weak predictors of *future* performance at almost all levels of the education/training system, and in working life afterwards.

This has long been the case with respect to performance in the education system. Thus, Fulton and Elwood (1989), for the Training Agency, pointed out that the usual criterion for access to higher education – A-level results – has very low predictive power with respect to future performance within it. In one large-scale study (Sear, 1983; cf. Leitch 1990) it was found that A-level scores accounted for less than 10 per cent of the variance in university performance. A similar pattern was found among polytechnic students (Rees 1981). A recent meta-analysis of over twenty studies confirms this general picture (Peers and Johnston 1994).

It is perhaps not surprising, in view of this, that many studies have shown that there is little relationship between academic 'potential' and performance in the workplace at any level, even, ironically enough, for future academics. For example, a meta-analysis by Cook (1988) reported correlations of around 0.1 between academic achievement and subsequent job performance. And (as seen in Chapter 4) there is little if any association between so-called 'measures of potential', like IQ, and actual job performance. Recent reviews in the USA (where a similar metaphor of potential prevails) report 'little or no relationship between academic and practical tasks' in a wide variety of unskilled, professional and managerial domains (Wagner 1994: 146).

Of course, those who are dismayed by such outcomes immediately suggest the involvement of 'other factors' (see, for example, the explanation given by Raven, Raven, and Court, 1993, for the failure of Raven's IQ scores to predict occupational performance). Which is precisely the point: there is barely a shred of objective evidence for the presence, or hidden residue, of a constant potential at work in people's cognitive development. Instead of supporting the notion of potential, an objective look at its 'flowering' suggests a phantom or spirit power rather than a substantive 'thing'. This was the view put many years ago by David McClelland (1973: 2), when he warned that the whole idea functions simply by perpetuating 'a mythological meritocracy in which none of the measures of merit bears significant demonstrable validity with respect to any measure outside of the charmed circle'.

One reason for clinging to the idea is, of course, that the institutions, such as the family and school, do not operate in isolation, but in the context of a whole socio-economic system that creates more general imperatives than the development of the abilities of all our children. As Ortner (1984: 154; cf. Lave 1994) points out, everyday practices 'embody within themselves the fundamental notions of temporal, spatial and social ordering that underlie and organise the [socio-cultural] system as a whole'. In other words, learning and teaching processes embody ideas which reflect the imperatives of the whole socio-cultural system in which we live. In our current market-oriented culture, this imperative involves the use of schools, above all else, in the identification and cognitive development of children who are going to 'contribute' to the economy and help us achieve domination in international competition. The idea of potential (and the selectivity it implies) legitimates an economic focusing of educational resources on the relatively few, and a rationalisation for the failure of the many. Teachers see this 'natural selection' function as a highly important and, indeed, 'professional', contribution to social and national goals. In fact, the whole gearing of the school and education system to a hierarchical job market makes it difficult for teachers to conceive of their role in any other way. Research shows quite clearly, though, that it is being achieved by a cyclical process of attribution and pseudo-assessment rather than the identification of cognitive potential as such.

Potential or expectations?

Very soon after entering school, at five years, children become scrutinised for 'signs' of their congenital potential for learning, and just as quickly become labelled accordingly (review in Rogers 1991). 'Pygmalion' and other 'expectancy' (read 'attribution of innate potential') effects are now well-established research findings. Usually this potential is actually assessed on the basis of a number of social, linguistic, and personality characteristics, rather than learning ability as such. Even facial appearance and other aspects of physical self-presentation have been shown to play an important part in the assessment of children's intelligence (Langlois 1986). Labelling leads in the more unfortunate children to poor self-expectations, low self-esteem, poor self-concept in the school situation, low sense of control over their own destiny, and thus poor motivation or even strong alienation (Rogers 1991).

These effects on children's sense of personal cognitive efficacy only reflect those also produced in a myriad ways in the wider society. Albert Bandura and colleagues (for review see Bandura *et al.* 1996) have suggested how parental socio-economic status already involves *socially* inherited concepts of personal cognitive efficacy that are, in turn, passed on to their children, irrespective of actual abilities. There seems little doubt that this transmission of efficacy beliefs is due, at least in part, to the culture of fatalism and pessimism created by psychologists of cognitive ability through the instrumentation of IQ. Parents' sense of efficacy, and their consequent sense of their *children's* cognitive efficacy, has been shown to be a major factor in the latters' cognitive development. As Bandura *et al.* note,

> The findings of diverse lines of research reveal that efficacy beliefs exert considerable impact on human development and adaptation. . . . Such beliefs influence aspirations and strength of goal commitments, level of motivation and perseverance in the face of difficulties and setbacks, resilience to adversity, quality of analytical thinking, causal attributions for successes and failures, and vulnerability to stress and depression.
>
> (Bandura *et al.* 1996: 1206)

The implicit (and occasionally explicit) 'drip' of negative impression on children's consciousness of their own abilities does not leave them cognitively unaffected. A related line of research has identified distinct patterns of emotion, cognition, and performance among such children in the face of challenging tasks: they display 'helplessness' when other children, not so affected, display 'mastery-orientation' (Smiley and Dweck 1994). 'Helpless' children experience negative feelings, have poor self-efficacy beliefs, underestimate their past performance, predict poor future performance, and thus actually perform poorer. In the same situation mastery-oriented children have more positive feelings, galvanise their cognitive self-efficacy, focus on ways of improving performance, and actually perform better in future. These patterns are unrelated to true ability on the tasks studied, as assessed by other means (Smiley and Dweck 1994).

Ability or persistence?

The problem is that the teaching–learning process in schools appears to be set up as a 'test' of such sense of personal efficacy under the guise of an objective 'test' of children's general learning potential. Researchers have noted this immediately from the way that knowledge to be learned is specially prepared with special characteristics. Designed both to be of some obscure future use and as a way of sorting out children on the basis of their basic learning potential, it boils down to a test, above all else, of their motivation and perseverance.

For one thing, school knowledge seems to be shorn of meaningful reference to the environments, economics, and social structures of children's communities that might help their development as responsible citizens. And it is abstracted, refined, and packaged in a way which seems to minimise its social and historical interest. Perret-Clermont and Bell (1988: 276–277) illustrate this with the subject of sets in school mathematics. They point out how set theory as elaborated by mathematicians becomes transformed by curriculum experts as an 'object' of the school curriculum.

> This transformed (deformed?) object finally reaches the child who in turn re-contextualises the notion of 'sets' as an object to be learned. Thus, at the end of this process the (naive) psychologist discovers that there is little in common between the notion of 'set' as initially elaborated by the mathematician and the notion of 'set' as learnt by pupils.
>
> (Perret-Clermont and Bell 1988: 276–277)

Further research has drawn attention to how this specially repackaged knowledge becomes more a test of children's perseverance and self-confidence than actual cognitive ability. There is serious disquiet among those who look carefully at what, indeed, children are learning in schools. Something 'sticks', of course, and there is little doubt that, by dint of teachers' hard work, children are given hundreds of experiences and activities they would not (under current social arrangements) otherwise have.

But there is also little doubt that it is a haphazard process with very uncertain outcomes. In schools, as Edwards and Mercer (1987) observed, most of the subject knowledge acquired consists of 'ritual knowledge', and what children are *really* learning are various 'discursive devices' through which they try to keep the teacher happy. Cole (1990) suggests that school learning involves, in the main, large amounts of fragmented information to be committed to memory; basic communication and computation (which many children, in other contexts, learn anyway – see below); and certain forms of knowledge classification. Meighan (1997: 14) warns that 'the "squirrels and nuts" theory that underpins the National Curriculum, whereby young people bury endless chunks of information in their memories in case it might just come in useful some day, is obsolete'. Most school experience, in other words, seems to be set up to test children's perseverance and learning confidence, which, in turn, is a reflection of their sense of personal cognitive efficacy. It is often the case, in fact, that children use the latter as a way of 'copping out' from such tedious demands.

Potential and pseudo-achievement

The fact that what children are learning in schools has little effect on their *general* knowledgeability of, and cognition about, the real world, is shown in other ways. It has frequently been shown how A-level and university students, steeped in objectified curriculum knowledge, are totally incapable of translating it to, and thus understanding, real-life practical situations in corresponding domains. Howard Gardner describes the results of a large number of studies on both sides of the Atlantic as follows:

> Perhaps most stunning is the case of physics. . . . (S)tudents who receive honours grades in college level physics are frequently unable to solve basic problems and questions encountered in a form slightly different from that on which they have been formally instructed and tested. . . . Indeed, in dozens of studies of this sort, young adults trained in science continue to exhibit the very same misconceptions and misunderstandings that one encounters in primary school children . . . essentially the same situation has been encountered in every scholastic domain in which inquiries have been conducted.
>
> (Gardner 1991: 3–4)

All this suggests that, although 'standards' remains a perennial issue in the education system, we clearly need to ask 'standards in *what*?' But the implications are both wider and deeper than this. Not only are pupils and students failing to learn in any deeper sense the knowledge that schools put before them, they are also *not* being given the chance of learning what they most need to in order to become responsible citizens. Thus, in a report of research for the Economic and Social Research Council, McGurk concluded that

> The overwhelming majority of British youth appear to be politically illiterate. They have no conception of the structure of society, of how the economy works, of the characteristics of

> different political systems; and they are hardly aware of the policy issues, let alone the philosophical differences, which distinguish the principal British political parties.
>
> (McGurk 1987: 6)

I will discuss these special aspects of the teaching–learning process in the education system in greater detail after I have considered again the 'cognition-culture complex' as the real basis of human learning. But it should already be clear that the system, based on the notion of innate potential, and all the quasi-biological and psychological assumptions underlying it, seems to be grossly dysfunctional to the needs of most children as well as the needs of society. Those who become certified to 'know' do not, apparently, know all that much; those who do not know it are said not to because they lack potential; and all our young seem debarred from learning what they most need to learn in order to be responsible citizens. Finally, as a result of their experiences in the education system, a large proportion of our young actually *believe* they do not have any learning potential and cannot succeed (see Stevenson and Palmer 1994: 184).

This is the bleak legacy of a system whose fundamental guiding principle is the cultivation of congenital potential. It is, of course, a grossly wasteful process in every sense of the word and a serious threat to any nation pretending to uphold a democracy. Just as the diseased, the hungry, and the homeless are in a very poor position to seize or create opportunities, plan productive activities, be self-reliant, and so on, so are those whose conceptions of their own cognitive ability have been damaged by erroneous assumptions and the criteria of ability that flow from them.

Potential found

This book has tried to put an alternative view of potential. In Chapter 2, I tried to show that, even at the genetic level, potential is *created* by more general genomic regulations 'guided' by conditions inside and outside the body. Then, in Chapter 3, I tried to show how genomic regulations have themselves become nested in other, epigenetic, regulations, and these in cognitive regulations, adaptable throughout life. Finally, I tried to show how humans evolved a further level of regulations, which dominate all others, and have made humans such a unique species. Humans act and produce, cooperate and communicate through cultural tools. When permitted to develop and operate with integrity these regulate all human behaviour, conditioning (indeed, in a sense, actually defining) all other needs: under socio-cognitive regulations, even the need for food can be denied to the point of death. Moreover, institutionalised in the guise of science and technology, social regulations provide means for intervening in and altering the regulations at all other levels, including learning and epigenetic and genetic regulations.

The essential point in this model, though, is that potentials are created in the interactions between these levels of regulations in relation to events in the outside world. They are not ones prefigured in genes (or environments for that matter). This is most clearly, and simply, seen in epigenetic interactions in which endpoints are, as it were, deliberately *not* predetermined in genes (for example, the bent versus the straight forms of barnacles, described in Chapter 3). Epigenetic regulations have 'opened up' genetic regulations, making them more adaptable to unpredictable circumstances. In subsequent evolutions epigenetic regulations have themselves been opened up to 'service' regulations at the higher physiological and cognitive levels, on a lifelong basis. New potentials for adaptation have been found in the interaction between the cognitive and epigenetic regulations. This arrangement has made organisms vastly more adaptable to unpredictable circumstances, so much so that the environment can be acted upon and changed in anticipation of future needs.

In humans, however, this creativity of potential lies in the relations between the cognitive regulations, on the one hand, and the social regulations in which they are embedded, on the other. Just as cognitive regulations transform the nature of epigenetic regulations, turning them (chiefly in the nervous system) into lifelong adaptabilities, so cultural regulations fundamentally alter the nature of human cognitions, compared with those of other animals. Within themselves, neither of these levels of regulation consists of static structures or processes (e.g. knowledge rules, stereotyped cognitions or behaviours). Whereas psychologists have long been preoccupied with describing the purely cognitive mechanisms which foster human social life, it is actually the latter which fosters the former: 'if we removed human activity from the system of social relationships, it would not exist . . . the human individual's activity is a system in the system of social relations. It does not exist without those relations' (Leontiev 1981: 46–47, quoted by Cole 1990: 91–92).

Let me briefly review some of the enormous benefits of this arrangement.

Productivity

The most conspicuous aspect of the results of this interactive relationship is, indeed, its productivity when operating in ideal ('free') conditions. Newell (1990: 116) calls it the 'efflorescence of adaptation':

> The fact is that [humans] seem to create new response functions all the time. To adopt the well-worn device of the Martian biologist studying the human species, what would impress it most is the efflorescence of adaptation. Humans appear to to go

> around simply creating opportunities of all kinds to build different response functions.
>
> (Newell 1990: 116)

Among these he instances jobs and occupations, games, cooking recipes, book-writing, dancing, music, rapping, building, and so forth. He then compares the creativity of humans with that of other animals and their singular, fixed adaptations.

> Each adaptation is seen as a unique biological phenomenon, to be curated one by one. Each is to be understood by exploring the behavioural and physiological mechanisms that support it. Not so with humans. There is no enumerating their adaptations – they will invent new adaptations faster than they can be recorded.
>
> (Newell 1990: 116)

Understanding this process has been the central problem of cognitive psychology. 'It is the dilemma of psychology to deal as a natural science with an object that creates history' (Cole 1990, quoting Ernst Boesch).

Newell (1990: 116) goes on to suggest that 'Indeed, our Martian biologist would not be wrong to conclude that the biggest biological puzzle about earthlings is why they have developed this efflorescence of adaptation'. I have tried to offer some solution to that puzzle in previous chapters. In the process I attempted to explain why it is a great mistake to reduce adaptation to the level of genetic potential and/or environmental determinism. To do so simply by-passes the creative richness of the human socio-cognitive system, and invites the kind of mysticism about 'gene gods' and good or bad environments that has so confused nature–nurture debates. It exists as a distinctive self-organising system.

Action with objects

Consider even the simplest behaviour of humans around objects. In humans (unlike cases with other animals) this virtually never consists of isolated cognitions around isolated entities. Rather, the objects mediate whole patterns of social relationships which individual cognitions 'enable'.

> In acting with objects the child is not merely learning the physical properties of things but mastering the social modes of acting with those things. These socially evolved modes of action are not inscribed in the objects themselves and cannot be discovered independently by the child from their physical properties – they

> must be learned through a socially-mediated process. . . . In embryonic form, the infant's relationship with others is mediated through objects – a bottle, a hand, a spoon. . . . As the child acquires and internalises speech, social interaction is mediated by powerful systems of symbolic communication which is capable of liberating activities from the restrictions of the settings in the here and now. The central point is that the child–object and the child–other interactions are part of a unitary process of activity for some purpose.
>
> (Scribner 1997: 268)

Vygotsky (e.g. 1981) made similar points about the nature of tools, both technical and psychological (see Chapter 3). Just as the technical tool transforms mental operations of those using them, so the psychological tool radically alters and extends what would be otherwise limited capacities. In this interaction new designs and forms are regularly conceived.

Concepts

One of the most spectacular examples of the creative potential of such a system is the construction of generalisations over sets or domains of objects and events – i.e. what are usually referred to as *concepts*, the mental representations of categories of external objects and events. The genesis of such concepts would not be possible without social interaction: indeed, they wholly presuppose one another.

> In order to transmit some experience or content of consciousness to another person, there is no other path than to ascribe the content to a known class, a known group of phenomena, and as we know this necessarily requires generalisation. Thus it turns out that social interaction necessarily presupposes generalisation and the development of word meaning, i.e. generalisation becomes possible with the development of social interaction. Thus, higher, uniquely human forms of psychological social interaction are possible only because human thinking reflects reality in a generalised way.
>
> (Vygotsky 1956: 51; cf. Semin 1990)

In generalisation social regulations radically alter cognitive regulations and the process unleashes enormous cognitive powers in people. It does this by acting as a bridge between the past experience of the culture and the creation of novel solutions in the novel particular. Yet all children display it as a common ability from a very early age (see Walsh, Richardson, and Faulkner 1993, for review). Socially cultivated and institutionalised in the form of scientific theory, such generalisation has been extraordinarily creative. Yet few if any philosophers of science today would argue that the new human potentials delivered by science have arisen from the machinations of isolated cognitions, detached from a social and historical context. Indeed, the creative, dialectical, development of scientific cognition in social context has recently become the intense focus of 'discourse analysis' or 'discursive constructivism' (Gergen and Semin 1990).

Maths

A similar cognitive creativity seems to apply when social contexts require thinking about quantitative relations. In contrast with the turgidity of maths learning in schools, studies of everyday mathematics (e.g. Lave 1988), even among ten-year-old street traders in Brazil (Nunes *et al.* 1993), show that 'quantitative relations are assembled inventively and effectively in everyday situations, independently of problem solvers' past school biographies' (Lave 1994: 318).

Memory

Or consider the human use of memory. The strongest tradition of memory research has required subjects to learn and immediately recall lists of digits or nonsense syllables. The purpose of this immediacy and nonsensity has been to avoid the effects of background knowledge or use of learning strategies. Complex models of human memory have been based on these dull individualistic routines.

Yet they are the very opposite of the creative way that human memory is actually used. Very early in human social history, the demands of social cooperation and communication required counting and simple computation, as in the management and distribution of crops and animals. The interaction between social needs/regulations and cognitive ones led readily to the invention of simple tallies – auxiliary tools which, as Vygotsky and Luria (1993: 177) explain, vastly expand the memory function, transforming the natural function in the process: 'It is this transition from natural forms of memory to the cultural ones that constitutes development of memory from child to adult'.

Reflective abstraction and science

As another example, we may consider 'reflective abstraction', the process by which new conceptions actually arise from the analogies/contradictions in existing conceptions, and/or between these and present experience. As both Piaget (e.g. 1965; cf. Doise 1988) and Vygotsky (1988) have pointed out, this important process, seen in only the most rudimentary form in other species, arises chiefly in the need to coordinate the conflicting points of view that are present in any situation of social cooperation. Here, again, the social form of action and experience transforms individual cognitive regulations, producing new forms of knowledge and reasoning, and still newer potentials.

It is in just this kind of interaction between the cognitive and the social that modern science – systematisation of observation, theory construction, hypothesis testing, or what Carey (1988) calls the 'logic of confirmation' – was founded. It has been pointed out how this logic of confirmation grew out of the upsurge in the *social* process of conferencing, cooperation and publication among scientists in the seventeenth to eighteenth centuries. Lieberman (1984), for example, claims that publishing results and scientific theories in journals that were accessible to large numbers of people was *the* crucial step in the development of the scientific age.

Piaget (e.g. 1965) has argued that pre-experimental (or Aristotelian) concepts of physics lie in relation to modern scientific ones much as the six-year-olds 'concrete operations' lie in relation to the 'formal operations' of mature cognition, and that the developmental change in both spheres reflects new levels of social interaction. The logic of confirmation, in other words, also appears to be a creative 'invention' of the interaction between individual cognitive and social regulations.

Socio-cognition and altruism

In Chapter 3 I tried to show how 'learning' (or, more accurately, lifelong cognitive development) consists of the construction of internal representations of the deep covariation structures in experience. Such representations make the world predictable, even on the basis of limited experience. For example, the blue-tit that attunes to a complex pattern of food distribution is doing so on the basis of covariations that lie deeper than superficial bivariate associations. Such representations, of course, include, as covariates, a number of internal appetitive and affective variables, such as hunger, the sensations of eating, of motor action, and so on. What really develop, as the basis of all intelligent behaviour, are cognitive-affective-motor 'schemas'.

In the cognition-culture complex of humans, though, the role, even of feelings, in intelligent behaviour becomes radically altered by incorporation into *socio-cognitive* schemas. Through the mediation of shared schemas we participate in an *interpersonal* cognitive structure that includes not only the thoughts and behaviours of others, but their feelings too. Through personal observations we can experience the pains and sufferings that others experience; through personal actions that benefit others, we actually share the pleasures and feelings that they enjoy This is the basis, I suggest, of the remarkable altruism of humans. Although the biological reductionism of the times has attempted to rationalise such altruism in terms of the subconscious ruminations of 'selfish genes' (Dawkins 1976), there is in fact a far richer *psychological* explanation, and it amounts to yet another aspect of the creativity of the cognition-culture complex.

In everyday contexts like the workplace, such cognitive inventiveness is almost always unrecognised by psychologists. In one study, Scribner (1997a) studied the quantitative cognitive strategies of dairy workers in, for example, the procurements of items from stock for orders, stocktaking, and the pricing of deliveries by delivery personnel. As with most established workplaces, the fundamental context is a set of social regulations in the form of evolved shared concepts, work procedures, and instruments. But workers did not do their myriad computations in this context by a mechanical application of standard mathematical algorithms. Rather, they exhibited a vast adaptability and variability of problem-solving, each improving speed and efficiency in meeting the wide range of specific orders and delivery prices.

> To work in such a coordinated and efficient manner, each assembler needed to have some internally represented knowledge of the spatial arrangement of the warehouse that could be used flexibly to organise the items on hand. . . . With such knowledge (and, of course, much more) they elegantly mapped one organisation on to the other – the symbolic on to the spatial – to meet their own needs and to satisfy externally imposed task requirements. Product assemblers creatively synthesized several domains of knowledge as a means of organising and regulating their own actions. . . . If social knowledge organises the dairy – its physical environment and symbolic forms – individuals use this knowledge creatively to shape work that is better adapted to to human needs.
> (Scribner 1997b: 298)

The consequences of splitting the cognition-culture complex

Scribner's (1997b: 298) final comment is that 'We can only regret that our social institutions . . . are so organised as to limit the ways in which the thought and action of individual workers can turn back, enrich and humanise social knowledge and practice'. What we see in the typical organisation of workplaces, though, is the splitting of the culture-cognition complex that so suppresses human potential in all our institutions.

Indeed, it suggests why all attempts to understand human cognitive potential shorn from its social context are doomed to failure and serve only to perpetuate mysticism. This is seen particularly in IQ tests. For example, the administration of a non-verbal test like the Raven's is considered to be a test of 'pure' reason, close to the legendary, but mythical, 'core' of intelligence called '*g*' (Carpenter, Just, and Shell 1990). Scores are supposed to locate children and others on a normal distribution of strength or power of '*g*' as a fixed potential, the distribution of which parallels that of favourable and unfavourable genes. On the one hand, it overlooks the (only slightly covert) culture-specific origins of Raven's items: indeed, even a cursory inspection of items (see Chapter 4 and Richardson 1996) easily exposes the myth that we are somehow testing cognition detached from social context and the social histories of children. But more importantly, it creates the unseemly confusion regularly found among psychologists whenever they are asked what this '*g*' or 'intelligence' actually is (Sternberg and Berg 1986; see Chapter 4).

I have already commented on how problems which arise from cleaving the cognition-culture complex are writ large in schools, and I intend to elaborate on those in the rest of this section. The reason that so many children appear to fail in schools is that the usual curriculum systematically and deliberately separates cognition from culture, as if putting them together would somehow detract from the very purpose of the education system. In school, knowledge and certain skills become the main 'objects' of activity. But these are not objects around which children can act and 'turn back on' freely: children do not 'own' them, as Lave (1994) put it. School knowledge is not 'knowledge' as we know it in the everyday sense – as the knowledge we use in work and social intercourse all the time. Nor is it 'knowledge' as we know it in the academic or scholarly sense, but a specially packaged variety. Its learning occurs as disjointed fragments and is motivated, on the part of pupils, by long-term status goals rather than current intrinsic interest; and, on the part of teachers (partly), as a means for sorting children out into the 'bright' and 'dull'. Its chief cognitive effects are in committing huge amounts of fragmentary information to memory and in knowledge classification (Cole 1990).

Because cognition and culture are split, skills, too, appear to be learned in schools in only a shallow sense, as much research in classrooms has recently shown. In the case of mathematics, for example, it seems true that pupils learn, laboriously, computational procedures and some rules for manipulating the symbols used in the mathematical notation system (Resnick 1986; cf. Lave 1994). But they fail to learn the deeper meanings of the symbols, in what ways they represent quantities and their transformations – representations and cognitions which only develop from their acquisition and application in real contexts. Children in the typical classroom, 'do not bring their mathematical resources to bear, have no control over the process and believe themselves unable to invent or discover procedures (because they are not mathematical geniuses)' (Lave 1994: 320). As a result they end up believing that mathematics is an especially difficult kind of knowledge and cognitive skill, instead of a form of social

discourse. Mathematics thus remains a 'dead' subject for the vast majority of school students instead of, potentially, an exciting and interesting tool. Much the same can be said about other skills such as reading, writing, graphic art, 'doing' science, and so on. Moreover, as Lave (1994) points out, the decompositions of knowledge and skills into more finely prescribed curricula and learning objectives, which is the current trend, will only make matters far worse.

Much the same has been observed with other areas of knowledge and skills, such as science. In consequence of a host of studies, educators are now stressing that, when children do manage to make sense of scientific concepts and definitions, it is because they can be related to everyday contexts (Moll 1990: 10). More generally, as Glaser (1984: 8) pointed out:

> learning and reasoning skills develop not as abstract mechanisms of heuristic search and memory processing. Rather they develop as the content and concepts of a knowledge domain are attained in learning situations that constrain this knowledge to serve certain goals and purposes.

Effective thinking is the result of 'conditionalised knowledge – knowledge that becomes associated with the conditions and constraints of its use'.

This pinpoints the central dilemma of schools, and increasing numbers of researchers and theorists have drawn attention to it in recent years: knowledge and skills are difficult to learn in school, or are learned in only a moribund form, because they are not part of real life. Indeed, as Cole (1990) points out, it is part of the distinctive remit of schools to equip children with the knowledge and skills they will need in later life *by* removing them (children, knowledge, and skills) from the contexts of practical activity. As mentioned above, most of what children are really learning in schools consists of various 'discursive devices' (Edwards and Mercer 1985) and/or confirmation of their cognitive self-efficacy, itself derived from their socio-economic status (Bandura *et al.* 1996).

Accordingly, school persists as part of real life in another sense, i.e. that of social role differentiation based on 'innate' potentials. There have been many studies of the social dynamics of that process. What Cole calls 'the participant structure' of schooling involves a special kind of discourse, the central goals of which are to set graded learning tasks to children that, in turn, provide the teacher with information about the quality, not only of the learner's progress, but of his/her general learning potential. As Cole (1990: 105; citing a study by Mehan 1979) points out,

> The central goals of instructional discourse are to give children information about the content of the curriculum and feedback

> about their efforts while providing teachers with feedback about the students' progress . . . (indeed) when the teacher asks a question it is really a covert way to evaluate the student's progress.
>
> (Cole 1990: 105)

In that process pupils become attributed with innate potentials for learning particular subjects, or with more or less general 'intelligence'.

When the cognition-culture complex is split, learning becomes laborious in the extreme. The consequences are observed regularly in the poor success rates for remedial or compensatory programmes for individual learners. Children at risk of failure in school remain remarkably refractory to these. Thus, in attempts to improve reading comprehension, for example, Stothard (1994: 230) argues that 'considering the immense effort devoted to teaching comprehension, the improvements produced have typically been surprisingly small'. She quotes Pearson and Gallagher (1983) on a more general point: 'as one looks across these various attempts at removing the roadblocks of knowledge deficits, what is impressive with very few exceptions, is how weak rather than how strong the effects are'. The APA group of psychologists makes similar points (Neisser *et al.* 1996).

More tragic is how such failures are sometimes turned around to 'blame the victim'. It is well known how, after reviewing the failure of compensatory education programmes in the USA, Jensen (1969) returned to 'genetic' explanations. Since most of the subjects were blacks this swiftly led to another round of scientific racism, including calls for segregation, eugenic measures, and so on (see contributions in Richardson and Spears 1972). Such calls have been echoed more recently in debates surrounding works by Herrnstein and Murray (1994) and others. As mentioned in Chapter 4, the very idea of 'race' and racism are immanent in the prevailing metaphysical model of potential, especially its attempted realisation in the IQ test, and the quasi-biological assumptions on which it is based.

Reuniting the cognition-culture complex

From the point of view of educational effectiveness, many researchers and educational theorists have stressed the need to put the cognition-culture complex back together in our schools as a most urgent problem. As Bruner (1985: 29) puts it, 'The fact that we learn the culture as readily and effectively as we do must give us pause – considering how poorly we do at certain artificial, "madeup" subjects that we teach in schools and whose use is *not* imbedded in any established cultural practice'. Wood (1991: 112) argues that the conditions for true cultural learning 'are more likely to be endemic to the home or local culture in a way that they are not to the school'. And Tharpe and Gallimore (1991) wonder why the ease of learning we find in cultural contexts is so conspicuously absent in schools. But as Wood goes on, such possibilities are

> tightly constrained by the nature of the institutions we have invented to bring teachers and learners together. If we find ourselves dissatisfied with the interactions that take place in such institutions, measured against what we take to be the optimum contexts for learning, then we must question not simply the teacher's 'skills' but the form of the institution within which we expect these to be deployed.
>
> (Wood 1991: 118)

These concerns are urgent and are the subject of much current discussion (e.g. Moll 1990). Many attempts are currently being made to break the mould (Rogoff 1994; Rogoff *et al.* 1995). But the problems run extremely deep for the simple reason that the splitting of the cognition-culture complex is the very *means* by which we have come to organise our economies and societies. By this I mean that most people are engaged in productive activity merely as labour units, divorced from control over the means and ends of that activity, with consequences, not only for the perpetuation of social-class structures, but for the psychological malaise of many individuals. Little wonder that industrialists like Charles Handy (1994) can speak of an urgent need for the 'cognitive enfranchising' of people at all levels of commerce and production. Instead of labour units, 'We ought, instead, to think of "membership" . . . membership gives meaning, and responsibility, to those who work in the business. They cease to be instruments or employees and become enfranchised' (Handy 1994: 42).

Of course, that schools need to work in the cognition-culture complex, rather than split it, has long been demanded by some educationists. But how they are to do it is another matter. Carraher and Schliemann (1988: 184–185) illustrate the issue with the subject of mathematics. They point out (as I also noted earlier) that the emphasis in school mathematics focuses upon the computational rules, and seldom on the *meaning* of the process. In contrast, informal mathematics (as in young street traders) preserves meaning and is used and expressed in a diversity of ways for different problems. The mathematics knowledge may thus be quite different across the two situations, and so transfer to novel problems with different ease. Schooled subjects are coached in rules, but 'the search for rules can be a source of improved performance when meaning plays a more important role' (Carraher and Schlieman 1988: 193). A teaching approach that combines both aspects would obviously approximate, rather more closely, mathematics knowledge as it has evolved among mathematicians, and is used in practical goals.

How could this possibly be done more generally? How could schools and communities be induced to 'reconnect' in a cognition-culture complex? The more we look at radical solutions to this problem, the more we soon come to realise the extent to which the educational system and its curriculum are ensnared in a whole system of production and ideology that limits change. Consider, for example, the following proposals. Schools may reconnect with communities by inviting, from the practical contexts of local producers and practitioners, the submission of genuine problems requiring thought, knowledge research, and practical action within the detached context of a school or college. For instance, the newsagent may have a delivery organisation problem, the Parish Council a reporting problem, the engineer a component design problem, the steelworks a marketing problem, the health centre a health education problem, the shirt factory another kind of design problem, the farmer all kinds of botanical and zoological problems, and so on, and so forth.

Within such real-life problems all the aims and objectives of any accepted curriculum – the development of skills of literacy and numeracy, of literature and scientific research, of computer-use, local and national history, geography, physics, biology, design, commerce, and so on – could be worked out. But they would be worked out in meaningful contexts which would not only help develop abstract concepts in a grounded way, but also engender economic sense, a sense of worthwhileness about activities in schools, as well as civic identity and responsibility. Above all, this approach would avoid the semi-enforced digestion of prepackaged 'dead' skills and knowledge, which turns school into the race of motivation and persistence which does so much damage to people's real potentials.

But such a scheme requires a radical switch of social and political priorities. Community producers and practitioners would have to be prepared to give up their time and concentration (though it may well benefit them in the long run) to identify and describe the problem, and talk to pupils or students at frequent intervals in schools and in the workplace. The latter, then, would have to be made safe and secure for children to visit and move around in. Workplace managers would have to take a very real and serious interest in education locally, and full-time liaison personnel may have to be recruited to arrange all that. Teachers then would have the far more challenging task of scaling the problem to the suitable level, identifying curriculum objectives within them, and then organising the literature and other research resources, the language-use, computations, practical action, and so on, to achieve them. This, too, would require the willingness on the part of parents (*all* parents) and others in the community to assist in this organisation, as well as the provision of time off work to do so.

I am naturally aware of such happenings (e.g. Industry-School links) in some schools already, but these arise only as a novel 'extra' on the far fringe of the curriculum. In contrast, I am suggesting that they should be central and systematic. Such an approach seems unlikely so long as the schools conceive their function as that of a search for potential through the mediation of decontextualised 'learning' routines. However, the need for such a reconceptualisation of practice seems an increasingly urgent one. With the accelerating globalisation of capitalist economics, the same principles of potential and learning and 'education' are now being adopted or more or less imposed on developing countries worldwide (see Cole 1990). The danger is that millions more children are in jeopardy of becoming ensnared in the same culture of pessimism that will suppress the potentials of the majority more definitely than ever before.

Conclusion

Theories of innate abilities – that we are preformed to think in predetermined ways with predetermined kinds of knowledge structures – do not make sense, either biologically or psychologically. And there is little if any respectable scientific evidence that individuals and groups who vary in their cognitive abilities do so because of their genes. Human potential is not 'in' our genes; nor is it 'in' our environments: rather, it is created by a developmental regulatory system, often, as Newell (1990) notes, with dazzling fecundity, when the conditions in the cognition-culture complex permit or require it. Human history is, of course, replete with cases in which attributed innate abilities have been overcome by overthrowing or lifting the ideology that defines them. I need only mention the recoveries from slavery; the sudden discovery of abilities of women to run economic production in times of war; and advances of working-class organisations in times of extreme oppression. Indeed, without potential in precisely that form, there would, quite simply, have been no human history.

This is not to argue that 'anybody can do almost anything'. But it is to stress that people's current potentials and cognitive statuses are products of *development* in a very broad sense, involving several levels of regulation in interaction and transaction with each other, not ones cast in DNA or environments. But neither are they cast in some static 'phenotypic' endpoint. There is inevitable diversity in developmental process and current statuses in humans. Not only must this be seen as valuable resource for the creation of future potentials in groups and individuals: there is an intrinsic indetermination in it, too. As Oyama (1985: 120) put it, 'The only way out of the problem of predetermined potential . . . is thus to see potential itself as having a history. It is multiplied, progressively determined, with new varieties of causes and consequences emerging at different hierarchical levels at different times'.

In the social processes by which they arise, human potentials, seemingly created by individuals, are actually the 'property', not of individuals, but of all of us, and ones in which we can all share. This is true in several senses. No sooner have new cultural tools become invented than they are rapidly assimilated by the rest. The same applies to the 'heights' of knowledge, as in scientific discovery. We justly praise the special efforts and achievements of individuals in the cognition-culture process, but most will readily acknowledge the role of that very context: ideas do not arise *de novo* in individual minds. As Einstein himself was always ready to point out, the previous theories of Maxwell and Lorentz 'led inevitably to the theory of relativity' (1934: 50), and insisted that the work of the individual is so bound up with that of scientific contemporaries that it appears almost as an impersonal product of the generation (Bergia 1979; Einstein 1982).

The 'putting back together' of the cognition-culture complex is of vital importance for the future potentials of humanity, but it depends on the thoughts and actions of other individuals in their organisations throughout the world. An author can do nought but help create a psychological climate in which any psychological doubts and uncertainties may be overcome and cultural progress encouraged. To psychologists I would say: Banish the spirits of potential that have so dogged our discipline. To parents and educators I would say: Impress upon children – *all* children – from the start, their immense powers as humans for learning; and then engage them in learning that seriously relates to their socio-economic milieu, and how to improve it. And to individuals I would say: Throw off your constraints and doubts about your own abilities – there is a whole world of human potential to be gained.

References

Bandura, A., Barbaranelli, C., Caprara, G.V., and Pastorelli, C. (1996). Multifaceted impact of self-efficacy beliefs on academic functioning. *Child Development*, 67, 1206–1222.

Bergia, S. (1979). Einstein and the birth of special relativity. In A.P. French (ed.) *Einstein: A Centenary Volume.* London: Heinemann.

Bruner, J.S. (1985). Vygotsky: a historical and contemporary perspective. In J.V. Wertsch (ed.) *Culture, Communication and Cognition: Vygotskian Perspectives.* Cambridge: Cambridge University Press.

Carey, S. (1988). Are children fundamentally different kinds of thinkers and learners than adults? In K. Richardson and S. Sheldon (eds) *Cognitive Development to Adolescence*. Hove: Erlbaum.

Carpenter, P.A., Just, M.A., and Shell, P. (1990). What one intelligence test measures: a theoretical account of the processing in the Raven Progressive Matrices Test. *Psychological Review*, 97, 404–431.

Carraher, T.N. and Schliemann, A.D. (1988). Culture, arithmetic and mathematical models. *Cultural Dynamics*, 1, 180–194.

Ceci, S.J. and Roazzi, A. (1994). The effects of context on cognition: postcards from Brazil. In R.J. Sternberg and R.K. Wagner (eds) *Mind in Context.* Cambridge: Cambridge University Press.

Cole, M. (1990). Cognitive development and formal schooling: the evidence from cross-cultural research. In L.C. Moll (ed.) *Vygotsky and Education: Instructional Implications and Applications of Sociohistorical Psychology.* Cambridge: Cambridge University Press.

Cook, M. (1988). *Personnel Selection and Productivity.* Chichester: Wiley.

Dawkins, R. (1989). *The Selfish Gene.* 2nd edn. Oxford: Oxford University Press.

Doise, W. (1988). Social development of intellect. In K. Richardson and S. Sheldon (eds) *Cognitive Development to Adolescence.* Hove: Erlbaum.

Edwards, D. and Mercer, N.M. (1987). *Common Knowledge.* London: Methuen.

Einstein, A.E. (1934). *The World as I See It.* New York: McCleod.

Einstein, A.E. (1982). *Ideas and Opinions* (compiled and ed. C. Seelig). New York: Crown Publications.

Ericsson, K.A. and Charness, N. (1994). Expert performance: its structure and acquisition. *American Psychologist*, 49, 725–747.

Feldman, D.H. (1986). *Nature's Gambit: Child Prodigies and the Development of Human Potential.* New York: Basic Books.

Fulton, O. and Elwood, S. (1989). *Admissions to Higher Education: Policies and Practice.* Sheffield: The Training Agency.

Gardner, H. (1991). *The Unschooled Mind.* London: Fontana.

George, J. and Glasgow, J. (1988). Street science and conventional science in the West Indies. *Studies in Science Education,* 15, 109–118.

Gergen, K.J. and Semin, G.R. (1990). Everyday understanding in science and daily life. In G.R. Semin and K.J. Gergen (eds) *Everyday Understanding: Social and Scientific Implications.* London: Sage.

Glaser, R. (1984). Education and thinking: the role of knowledge. *American Psychologist*, 39, 6–17.

Goodnow, J.J. (1984). On being judged 'intelligent'. In P.S. Fry (ed.) *Changing Conceptions of Intelligence and Intellectual Functioning.* Amsterdam: North-Holland.

Handy, C.B. (1994). *The Empty Raincoat.* London: Hutchinson.

Herrnstein, R.J. and Murray, C. (1994). *The Bell Curve.* New York: Free Press.

Howe, M.J.A. (1990). *The Origins of Exceptional Abilities.* Oxford: Blackwell.

Jensen, A.R. (1969). How much can we boost IQ and educational achievement? *Harvard Educational Review*, 39, 1–123.

Langlois, J.H. (1986). From the eye of the beholder to behavioral reality: development of social behaviors and social relations as a function of physical attractiveness. In C.P. Herman, M. Zanna, and E.T. Higgins (eds) *Physical Appearances, Stigma and Social Behaviour.* Hillsdale, NJ, Erlbaum.

Lave, J. (1988). *Cognition in practice: Mind, Mathematics and Culture in Everyday Life.* Cambridge: Cambridge University Press.

Lave, J. (1994). The culture of acquisition and the practice of understanding. In J.W. Stigler, R.H. Shweder, and G. Herdt (eds) *Cultural Psychology.* Cambridge: Cambridge University Press.

Leitch, A. (1990). *Factors Associated with Academic Failure: Their Role in Identifying Students at Risk of Failure.* Unpublished MSc dissertation: The Open University.

Leontiev, A.N. (1981). *Problems in the Development of Mind.* Moscow: Progress.

Lieberman, P. (1984). *The Biological Evolution of Language.* Cambridge, MA: Harvard University Press.

McClelland, D.C. (1973). Testing for competence rather than for 'intelligence'. *American Psychologist*, 28, 1–14.

McGurk, H. (ed.) (1987). *What Next?* London: Economic and Social Research Council.

Mehan, H. (1979). *Learning Lessons.* Cambridge, MA: Harvard University Press.

Meighan, R. (1997). Letter. *Guardian*, 22 January.

Moll, L.C. (1990). Introduction. In L.C. Moll (ed.) *Vygotsky and Education: Instructional Implications and Applications of Sociohistorical Psychology.* Cambridge: Cambridge University Press.

Neisser, U., Boodoo, G., Bouchard, T.J. Jr, Boykin, A.W., Brody, N., Ceci, S.J., Halpern, D.F., Loehlin, J.C., Perloff, R., Sternberg, R.J., and Urbina, S. (1996). Intelligence: knowns and unknowns. *American Psychologist*, 51, 77–101.

Newell, A. (1990). *Unified Theories of Cognition.* Cambridge, MA: Harvard University Press.

Nunes, T.M., Schliemann, A.D., and Carraher, D.W. (1993). *Street Mathematics and School Mathematics.* Cambridge: Cambridge University Press.

Ortner, S.B. (1984). Theory in anthropology since the sixties. *Comparative Studies in Society and History*, 26, 126–166.

Oyama, S. (1985). *The Ontogeny of Information.* Cambridge: Cambridge University Press.

Pearson, P.D. and Gallagher, M. (1983). The instruction of reading comprehension. *Contemporary Educational Psychology*, 8, 317–344.

Peers, I. and Johnston, M. (1994). Influence of learning context on the relationship between A-level attainment and final degree performance: a meta-analytic review. *British Journal of Educational Psychology*, 64, 1–18.

Perret-Clermont, A.-N. and Bell, N. (1988) Learning processes in social and instructional interactions. In K. Richardson and S. Sheldon (eds) *Cognitive Development to Adolescence,* Hove: Erlbaum.

Piaget, J. (1965). *Sociological Studies.* Geneva: Droz.

Raven, J., Raven, J.C. and Court, J.H. (1993). *Manual for Raven's Progressive Matrices and Vocabulary Scales: Section 1.* Oxford: Oxford Psychologists Press.

Rees, D. (1981). 'A' levels, age and degree performance. *Higher Education Review*, 13, 45–57.

Resnick, L.B. (1986). Constructing knowledge in school. In L.S. Liben and D.H. Feldman (eds) *Development and Learning: Conflict or Congruence.* Hillsdale, NJ: Erlbaum.

Richardson, K. (1996). Putting Raven into context. *British Journal of Educational Psychology*, 66, 535–538.

Richardson, K. and Spears, D. (eds) (1972). *Race, Culture and Intelligence.* Harmondsworth: Penguin Education.

Rogers, C. (1991). Early admission: early labelling. In M. Woodhead, P. Light, and R. Carr (eds) *Growing Up in a Changing Society*. London: Routledge in association with the Open University.

Rogoff, B. (1993). Children's guided participation and participatory appropriation in sociocultural activity. In R.H. Wozniak and K.W. Fischer (eds) *Development in Context: Acting and Thinking in Specific Environments*. Hove: Erlbaum.

Rogoff, B. (1994). Developing understanding of the idea of communities of learners. *Mind, Culture and Activity*, 1, 209–229.

Rogoff, B., Baker-Sennett, J., Lacasa, P., and Goldsmith, D. (1995). Development through participation in sociocultural activity. In J.J. Goodnow, P.J. Miller, and F. Kessel (eds) *Cultural Practices as Contexts for Development*. (New directions for child development No. 67). San Francisco: Jossey-Bass.

Sear, K. (1983). The correlation between A-level grades and degree results in England and Wales. *Higher Education*, 12, 609–619.

Scribner, S. (1997a). Mind in action: a functional approach to thinking. In E. Tobach, L.M.W. Martin, R.J. Falmagne, A.S. Scribner, and M.B. Parlee (eds) *Mind and Social Practice: Selected Writings of Sylvia Scribner*. Cambridge: Cambridge University Press.

Scribner, S. (1997b). Knowledge at work. In E. Tobach, L.M.W. Martin, R.J. Falmagne, A.S. Scribner, and M.B. Parlee (eds) *Mind and Social Practice: Selected Writings of Sylvia Scribner*. Cambridge: Cambridge University Press.

Semin, G.R. (1990). Everyday assumptions, language and personality. In G.R. Semin and K.J. Gergen (eds) *Everyday Understanding: Social and Scientific Implications*. London: Sage.

Smiley, P.A. and Dweck, C.S. (1994). Individual differences in achievement goals among young children. *Child Development*, 65, 1723–1743.

Sternberg, R.J. and Berg, C.A. (1986). Quantitative integration: definitions of intelligence: a comparison of the 1921 and 1986 symposia. In D.K. Detterman and R.J. Sternberg (eds) *What is Intelligence? Contemporary Viewpoints on its Nature and Definition*. Norwood, NJ: Ablex.

Stevenson, R. and Palmer, J. (1994). *Learning: Principles, Processes and Practices*. London: Cassell.

Stothard, S. (1994). The nature and treatment of reading comprehension. In C. Hulme and M. Snowling (eds) *Reading Development and Dyslexia*. London: Whurr.

Tharpe, R. and Gallimore, R. (1991). A theory of teaching as assisted performance. In P. Light, S. Sheldon, and M. Woodhead (eds) *Learning to Think*. London: Routledge in association with The Open University.

Vygotsky, L.S. (1956). *Selected Psychological Research*. Moscow: Izdatel'tsvo Akedemii Pedagocheskikh Nauk.

Vygotsky, L.S. (1988). The genesis of higher mental functions. In K. Richardson and S. Sheldon (eds) *Cognitive Development to Adolescence*. Hove: Erlbaum.

Vygotsky, L.S. and Luria, A.R. (1993). *Studies on the History of Behaviour* (ed. and trans. V.I. Golod and J.E. Knox). Hove: Erlbaum.

Wagner, R.K. (1994). Context counts: the case of cognitive ability testing for job selection. In R.J. Sternberg and R.K. Wagner (eds) *Mind in Context: Interactionist Perspectives on Human Intelligence*. Cambridge: Cambridge University Press.

Walsh, M., Richardson, K., and Faulkner, D. (1993). Perceptual, thematic and taxonomic relations in children's mental representations. *European Journal of Psychology of Education*, 8, 85–102.

Wood, D. (1991) Aspects of teaching and learning. In P. Light, S. Sheldon, and M. Woodhead (eds) *Learning to Think*. London: Routledge in association with The Open University.

NAME INDEX

SUBJECT INDEX

See also references to individuals should be checked in the Name Index.